Boston Studies in the Philosophy and History of Science

Volume 320

The series *Boston Studies in the Philosophy and History of Science* was conceived in the broadest framework of interdisciplinary and international concerns. Natural scientists, mathematicians, social scientists and philosophers have contributed to the series, as have historians and sociologists of science, linguists, psychologists, physicians, and literary critics.

The series has been able to include works by authors from many other countries around the world.

The editors believe that the history and philosophy of science should itself be scientific, self-consciously critical, humane as well as rational, sceptical and undogmatic while also receptive to discussion of first principles. One of the aims of Boston Studies, therefore, is to develop collaboration among scientists, historians and philosophers.

Boston Studies in the Philosophy and History of Science looks into and reflects on interactions between epistemological and historical dimensions in an effort to understand the scientific enterprise from every viewpoint.

More information about this series at http://www.springer.com/series/5710

Yafeng Shan

Doing Integrated History and Philosophy of Science: A Case Study of the Origin of Genetics

Springer

Yafeng Shan
Department of Philosophy
University of Kent
Canterbury, UK

ISSN 0068-0346 ISSN 2214-7942 (electronic)
Boston Studies in the Philosophy and History of Science
ISBN 978-3-030-50616-2 ISBN 978-3-030-50617-9 (eBook)
https://doi.org/10.1007/978-3-030-50617-9

This Springer imprint is published by the registered company Springer Nature Switzerland AG.
The registered company address is: Gewerbestrasse 11, 6330 Cham, Switzerland

Acknowledgments

Most chapters of this book were written during my IHPLS research fellowship (2018–2019) at the Cohn Institute for the History and Philosophy of Science and Ideas, Tel Aviv University. I am grateful to the Cohn Institute and Inter-University Programme in History and Philosophy of Life Sciences for the institutional and financial support. I would like to thank my colleagues at the Cohn Institute, especially Jose Brunner, Eva Jablonka, Shaul Katzir, and Ido Yavetz, and my fellows at IHPLS programme: Aya Evron, Topaz Halperin, Oren Harman, Arnon Levy, Jonathan Najenson, Nadav Rubinstein, and Ayelet Shavit. In particular, I would like to express my gratitude to Ehud Lamm and Adam Krashniak for their help during my time in Israel.

The idea of this book originated during my PhD study at University College London (UCL), while the book was completed at the University of Kent. I am very grateful to all those who have helped my thinking and writing. When I was a PhD student at UCL, I was strongly influenced and luckily nurtured by the integrated History and Philosophy of Science community in the UK. I am especially indebted to Hasok Chang and Greg Radick, who were my PhD examiners. Hasok's influence is enormous through his writings on integrated HPS and his sharp and constructive comments on my ideas. Greg's influence is also tremendous. Without Greg, Chap. 4 would have never existed. It is Greg who introduced me to Weldon's fascinating work. I am grateful to Jon Williamson for his invaluable advice and suggestions. I would like to thank Mike Buttolph for extensive discussion on Mendel and the history of genetics, his careful reading of the early drafts of the chapters in Part I, and giving me the copies of Bailey's paper and Darbirshire's paper. I would like to extend special thanks to Hugh Mackenzie for the helpful discussion and reading the chapters in Part II. I would also like to thank Jonathon Hricko for discussing the early versions of some arguments in this book.

Early versions of some arguments were presented at various conferences, including the meetings of International Society of History, Philosophy, and Social Studies of Biology (2013, 2015, 2017, 2019); the Philosophy of Biology in the UK conference (2016); the International Congress of Logic, Methodology, and Philosophy of Science (2015); the annual UK Integrated History and Philosophy of Science Workshop (2013, 2014, 2015); and various seminars and workshops in Durham, Helsinki, Jerusalem, London, Singapore, Sofia, Tel Aviv, and Zurich. I benefited much from the discussion with the audience there. In particular, I would like to thank Garland E. Allen, Theodore Arabatzis, L. A. Callender, Lindley Darden, Michael Friedman, Jonathan Hodge, Frank James, Samir Okasha, Nils Roll-Hansen, Sahotra Sarkar, and C. Kenneth Waters for their stimulating comments and encouragement.

I am also grateful to Rafa Siodor and his colleagues at UCL Special Collections for their help when I was archiving Weldon's manuscripts there. I would also like to thank two anonymous referees for Springer, for extremely constructive and helpful comments.

I would like to thank the great team at Springer who helped to bring this book to print. In particular, I would like to thank Lucy Fleet, Silvie Demandt, Prasad Gurunadham, Christopher Wilby, and S. Keerthi Kumari for their patience and efficiency.

I also owe my thanks and formal acknowledgment to publishers for granting me permission to revise and republish parts of my own previously published work as follows: Chapter 5 incorporates material from Shan, Yafeng. 2020. "Kuhn's 'Wrong Turning' and Legacy Today." *Synthese* 197 (1): 381–406. Chapter 6 incorporates material from Shan, Yafeng. 2019. "A New Functional Approach to Scientific Progress." *Philosophy of Science* 86 (4): 739–58.

Last but not least, I would like to thank my wife, Zifei, for her support and companionship. I would also like to thank Ajiao, Kai, San, and Wendi for their moral support.

This book is financially supported by Israel Science Foundation, British Society for the History of Science, and the Leverhulme Trust.

Contents

1 Introduction 1
1.1 History and Philosophy of Science 1
1.2 Integrated History and Philosophy of Science 4
1.3 Integrated HPS in Practice: The Case of the Origin of Genetics 7
References 9

Part I History

2 Mendel's *Pisum* Revisited 15
2.1 Mendel's Concern 15
2.2 Mendel and Gärtner on *Entwicklung* (Development) 16
2.3 Mendel's "*Entwicklungsreihe* (Developmental Series)" 24
2.4 Mendel's Novel Conceptualisation: The Laws of Developmental Series 27
2.5 Mendel and the Study of Heredity 30
2.6 Conclusion 33
References 34

3 De Vries' Mendelism Reassessed 37
3.1 The Rediscovery Story 37
3.2 No Mendel, No Mendelians! 39
3.2.1 The 3: 1 Ratio in the 1896 Notes 40
3.2.2 Mendel and the Law of Segregation 42
3.3 De Vries' Introduction of Segregation 44
3.3.1 From Mendel to Mendelism 44
3.3.2 From Activeness to Dominance 45
3.3.3 From Correspondence to Segregation 48
3.4 Conclusion 49
References 50

4 **Weldon's Choice Reconsidered** . . . 53
4.1 The Mendelian-Biometrician Controversy . . . 53
4.2 Weldon as a Biometrician . . . 54
4.3 Weldon's *Theory of Inheritance* . . . 56
4.3.1 The Outline . . . 57
4.3.2 The Methodology of the Study of Inheritance . . . 58
4.3.3 The Aim of *Theory of Inheritance* . . . 60
4.3.4 The Theory . . . 61
4.3.5 Summary and Remarks . . . 63
4.4 Weldon, No Biometrician? . . . 64
4.4.1 Pearson vs. Weldon Reconsidered . . . 64
4.4.2 Beyond Mendelism and Biometry . . . 66
4.5 Conclusion . . . 67
References . . . 67

Part II Integrated HPS

5 **Exemplarising the Origin of Genetics** . . . 73
5.1 Introduction . . . 73
5.2 The Theory-Based Accounts of the Origin of Genetics . . . 74
5.3 The Kuhnian Accounts . . . 80
5.4 A New Interpretation of Exemplar and the Exemplar-Based Approach . . . 81
5.5 An Exemplar-Based Account of the Origin of Genetics . . . 86
5.6 Conclusion . . . 96
References . . . 97

6 **A Functional Account of the Progress in Early Genetics** . . . 101
6.1 Scientific Progress and the Origin of Genetics . . . 101
6.2 A New Functional Approach to Scientific Progress . . . 102
6.3 How Early Genetics Progressed . . . 104
6.4 The Problems of the Kuhn-Laudan Functional Approach Revisited . . . 107
6.5 Beyond Knowledge, Truth, and Intervening . . . 111
6.6 Conclusion . . . 116
References . . . 116

7 **The Problem of the Long Neglect Revisited: An Exemplar-Based Explanation** . . . 119
7.1 Two Problems of the Long Neglect . . . 119
7.2 The Traditional Diagnoses of the Long Neglect . . . 121
7.2.1 Explanation 1: Mendel's Work Was Not Accepted . . . 121
7.2.2 Explanation 2: Mendel's Work Was Unknown . . . 123
7.2.3 Summary and Remarks . . . 124

7.3 Mendel's Contribution Reconsidered 125
7.3.1 The Traditional Philosophical Analyses 125
7.3.2 The Exemplar-Based Analysis 126
7.4 Why Mendel's Contribution Was Neglected 127
7.4.1 The Nature of the Long Neglect 127
7.4.2 The Exemplar-based Explanation 128
7.4.3 The Exemplar-Based Explanation and Old Intellectual Explanations 130
7.5 Conclusion 130
References 131

Part III Philosophy

8 A New Mode of Conceptual Continuity 137
8.1 Conceptual Change: Variance and Continuity 137
8.2 What if Everything Changes? The Case of the Concept of Dominance 140
8.3 A Holistic Approach to Conceptual Change 146
8.4 Two Modes of Continuity and Conceptual Continuity 148
8.5 The Case of the Concept of Dominance Revisited 151
8.6 Conclusion 154
References 155

9 The Gap Problem in Hypothetico-Deductivism 159
9.1 Mendel's Evidence and the Gap Problem 159
9.2 The Diagnosis of the Gap Problem and the Achinsteinian Solution 162
9.3 A New Solution: From Evidence to Evidential Practice 166
9.4 The Gap Problem Revisited 169
9.4.1 The Defence of Descriptive Adequacy 169
9.4.2 The Defence of Philosophical Adequacy 169
9.5 Normativity and Contextualism in H-D Evidential Practice 172
9.6 Conclusion 174
References 174

10 Promisingness in Theory Choice 177
10.1 Theory Choice in Science 177
10.2 The Choices in the Mendelian-Biometrician Controversy 178
10.3 Promisingness as Potential Usefulness 181
10.4 Promisingness, Potential Progressiveness, Potential Fertility, and Fruitfulness 184
10.5 Argument from Normativity 189
10.6 Conclusion 189
References 190

Appendix 193

Index 195

Chapter 1
Introduction

Philosophy of science without history of science is empty; history of science without philosophy of science is blind.

(Lakatos 1970, 91)

Abstract History and philosophy of science (HPS) has been controversial since its birth. In particular, the legitimacy of its status as an academic discipline has been constantly challenged and questioned. Nevertheless, substantial hope for HPS as a productive academic discipline is not dead. Some historically-minded philosophers of science and philosophically-minded historians of science never stop making efforts to promote the dialogue across the boundaries and defend HPS against the challenges in various ways. Integrated history and philosophy of science (integrated HPS) is such an attempt. It aims at a balanced inquiry from which both history of science and philosophy of science can benefit. In a nutshell, integrated HPS maintains that HPS should be both a good philosophy of science and a good history of science at the same time.

Keywords HPS · Integrated HPS · Anti-anachronistic reading

1.1 History and Philosophy of Science

In the late 1950s, more and more philosophers of science, including Norwood Russell Hanson, Mary Hesse, and Stephen Toulmin, recognised the significance of history of science in the philosophical examination of science. It does not mean that pre-1950 philosophy of science work is history-free. Rather in the 1950s philosophers of science became more careful and serious in using historical cases to philosophise. This tendency was greatly reinforced by the works of Thomas Kuhn, Imre Lakatos, Paul Feyerabend, Larry Laudan and others in the following decades. Consequently, as Ronald Giere (1973, 291) observed, "[T]hose philosophers of science who make serious use of the history of science form a loosely connected

Y. Shan, *Doing Integrated History and Philosophy of Science: A Case Study of the Origin of Genetics*, Boston Studies in the Philosophy and History of Science 320,
https://doi.org/10.1007/978-3-030-50617-9_1

school within the philosophy of science." This is what is later called "the historical turn" in the twentieth century philosophy of science (Bird 2008).

Concurrently, new postgraduate programmes and departments with an interdisciplinary focus on the historical and philosophical studies of science were founded in the United Kingdom and the United States. The Division of History and Philosophy of Science within the Department of Philosophy at University of Leeds was established in 1956,[1] while the Department of History and Philosophy of Science at Indiana University, the first department of its kind in the United States, was founded in 1960,[2] quickly followed by the establishment of similar departments in North America.[3] In 1970, *Studies in History and Philosophy of Science*, a new interdisciplinary journal that "devoted to the study of the conceptual history and foundations of science", was established by Gerd Buchdahl and Laudan. The historical turn, as well as the creation of new programmes, departments, and journals, advances the formation of history and philosophy of science (HPS) as a new academic discipline.

However, HPS has been controversial since its birth. In particular, the legitimacy of its status as an academic discipline has been constantly challenged and questioned. For modest sceptics, "[w]hether [a marriage of convenience between history of science and philosophy of science] will prove to be relatively permanent or only transitional remains to be seen" (Giere 1973, 296). For radical sceptics, "the failure of HPS" is simply a fact (Kuukkanen 2016, 3). To some extent, such scepticism seems to be justified in various ways. Institutionally, HPS is not very well developed in general, despite that there are still a number of HPS departments and postgraduate programmes worldwide. After a quick development in the 1970s, the institutionalisation of HPS slowed down. As Laudan (1989, 11) noted, "[HPS] slowly atrophied, and was finally officially pronounced dead at Princeton and Pennsylvania. It flickered briefly at numerous other places (e.g. Minnesota, MIT, Brandeis, Oxford, Oklahoma, Sussex, Melbourne, John Hopkins) but only occasionally became more than a rocky marriage of convenience at any of the latter." Thus, it is no wonder that Steven Shapin and Simon Schaffer (2011, xxi) regard the creation of HPS departments as "a largely unsuccessful experiment."

Axiologically, HPS is not easily viable, especially from a historian's viewpoint. It is commonly held that history of science and philosophy of science have distinct aims and problems (Giere 1973; Cohen 1977; Pinnick and Gale 2000; Steinle and

---

[1]Although it is often claimed that the Division of History and Philosophy of Science at Leeds is one of the oldest institutions of its kind in the world, the oldest HPS department in the UK was founded at University College London (UCL) much earlier. It was established in 1921 under the name "Department of History and Method of science", though it was renamed twice later ("Department of History and Philosophy of science" in 1938, and "Department of Science and Technology Studies" in 1994).

[2]Facing the resistance from the philosophy department, HPS department at Indiana was initially named "the department of history and logic of science" by Hanson (Grau 1999, S302).

[3]The HPS graduate program was launched at Princeton University in 1961 and at University of Pennsylvania in 1962. HPS department at Pittsburgh was established in 1970 while the HPS graduate program was introduced 1 year earlier.

Burian 2002; Kuukkanen 2016). And these differences are still growing. As Friedrich Steinle and Richard Burian (2002, 392) indicate, "there is a growing gulf between philosophical studies of conceptual change on the one hand, and cultural studies of scientific practices (which have recently helped to reshape the history of science) on the other." Today, historians of science tend to view science as a social phenomenon and thus focuses more on the issues of social histories, cultural histories, and histories of materials. In contrast, contemporary philosophers of science are fundamentally interested in the epistemic content of science. Accordingly, it is more and more difficult for philosophers of science and historians of science to find the works of each other that are relevant to their own concerns. In other words, it seems that "history and philosophy of science do not have much to say to one another." (Arabatzis and Schickore 2012, 396) Therefore, for many, the axiological parallel dooms a pessimistic prospect of HPS.

Methodologically, HPS has been heavily criticised. Most of the HPS works in the past four decades (e.g. the historicist approach to the scientific realism debate) can be characterised as the instances of the confrontation model, of which the basic idea is to test philosophical theories against historical data.[4] However, this confrontation model is shown to be problematic. Firstly, it implicitly assumes that there is an objective or neutral historiography which is capable of producing historical data that can be used in a straightforward manner to test philosophical theories. However, as Hans Radder (1997, 638) indicates, this assumption is highly implausible. It is doubtful that there is only one proper approach to the history of science, and even if so, more is to be said about which is the proper one and why. Secondly, the confrontation model oversimplifies the ways of understanding scientific practice. Jutta Schickore (2011, 471–74), for example, famously argues that the study of science, whether historical or philosophical, results from a hermeneutic analysis, "in which preliminary concepts and frameworks and initial case judgments are modified and adjusted until a cogent account is obtained, and this procedure should be reflected in our writing about science." Clearly, the confrontation model fails to follow this procedure. Thus, it is in this sense, Schickore (2011, 477) argues, that the confrontation model is "misleading."

Practically, the interaction between historians and philosophers of science is unbalanced or asymmetrical. Despite its name, HPS, as a product resulting from the historical turn in the twentieth century philosophy of science, has been fundamentally philosophically-oriented since the very beginning. Philosophers of science not only have contributed to the most of HPS work, but also have shown great interest in studying and writing the history of science. In contrast, many historians of science have been uninterested in philosophy of science and even sceptical of the

[4]Of course, there are alternative ways of analysing the methods of HPS. Some (e.g. Pinnick and Gale 2000; Scholl and Räz 2016) summarise the typical method of HPS as the case-study method, while others (e.g. Burian 2002) distinguish a top-down approach with a bottom-up approach in HPS. Nevertheless, I think that all of the case-study method, the top-down approach, and the bottom-up approach can be construed as instances of the confrontation model, since each follows the rationale that history serves evidence for philosophising.

interaction of history and philosophy of science. As Giere (2011) points out, "the presumption that what philosophers say is irrelevant to the work of historians of science seems still strong."

What is more, recent citation analyses also question that there is an academic discipline called HPS. According to K. Brad Wray's analysis (2010), no history of science journals is listed in the top journals cited in the major philosophy of science works. Wray (2010, 428), thus, argues that "if there were such a field as history and philosophy of science, one would expect scholars in that field to be citing publications in the leading history of science journal. But, it appears that philosophy of science is largely independent of the history of science."

1.2 Integrated History and Philosophy of Science

Nevertheless, substantial hope for HPS as a productive academic discipline is not dead. Some historically-minded philosophers of science and philosophically-minded historians of science never stop making efforts to promote the dialogue across the boundaries and defend HPS against the challenges in various ways. Institutionally, HPS conferences are regularly organised.[5] Axiologically, the necessity of history-philosophy engagement has been defended (e.g. Chang 2012a; Arabatzis 2017). A common ground has also been explored. For example, Michael Friedman (2008) and Alan Richardson (2008) actively promote the history of the philosophy of science (HOPOS) as a new variant of HPS which they believe may interest both historians of science and philosophers of science and "provides further ground for hope that a new kind of productive relationship between the two fields may now be possible." Methodologically, the confrontation model has been defended (e.g. Pinnick and Gale 2000; Scholl and Räz 2016; McAllister 2018; Scholl 2018). New HPS methods have been introduced and developed (e.g. Chang 2004, 2012a, b; Schickore 2009, 2011). In the recent development of HPS, I find a new movement called integrated history and philosophy of science (integrated HPS) most promising. Its manifesto is spelt out as follows.

> &HPS is distinctive in that it is both historical and philosophical at the same time.
>
> Good history and philosophy of science is not just history of science into which some philosophy of science may enter, or philosophy of science into which some history of science may enter. It is work that is both historical and philosophical at the same time.

[5]International conferences on integrated HPS take place every 2 years, while the UK integrated HPS workshop is held annually. Other major HPS-related conferences include the biennial meetings of International Society of History, Philosophy, and Social Studies of Biology (ISHPSSB), International Society of History of Philosophy of Science (HOPOS), and Society of Philosophy of Science in Practice (SPSP). In addition, there are regular HPS symposia in general philosophy of science conferences like the biennial meetings of Philosophy of Science Association (PSA) and the International Congress of Logic, Methodology, and Philosophy of Science (CLMPS).

> The founding insight of the modern discipline of HPS is that history and philosophy have a special affinity and one can effectively advance both simultaneously.
>
> What gives HPS its distinctive character is the conviction that the common goal of understanding of science can be pursued by dual, interdependent means. This duality may be localized in a single work. Or it may be distributed across many works and many scholars, with parts locally devoted just to historical or philosophical analysis. Intellectual history, for example, serves this purpose. What unifies this local scholarship into an HPS community is the broader expectation that all the work will ultimately contribute to the common goal.
>
> There is no distinct methodology that is HPS. Doing HPS does not confer a free pass to suspend the standards of one field to advance the other. It must be good history of science and philosophy, in that its claims are based on a solid grounding in appropriate sources and are located in the relevant context. And it must be good philosophy of science, in that it is cognizant of the literature in modern philosophy of science and its claims are, without compromise, articulated simply and clearly and supported by cogent argumentation. (Arabatzis and Howard 2015, 1–2)

The most distinctive feature of integrated HPS is integratedness. Unlike the other methodologies of HPS, integrated HPS is not overwhelmingly philosophically-driven. Too much attention has been paid to how philosophy of science can unbiasedly and wisely use history of science (see Burian 1977; Nickles 1995; Pinnick and Gale 2000; Scholl and Räz 2016; McAllister 2018; Scholl 2018). In contrast, integrated HPS aims at a more balanced inquiry from which both history of science and philosophy of science can benefit. Therefore, HPS should not be identical with a good historically-informed philosophy of science. Nor should it be merely a good philosophically-driven history of science. HPS should be both a good philosophy of science and a good history of science at the same time.

Another distinctive feature of integrated HPS is plurality. Axiologically, integrated HPS encourages a plurality of approaches to its aim. Methodologically, "let a hundred flower bloom" as long as a better history and philosophy of science is aimed at. There is no such a thing as the only correct method of integrated HPS.

This book defends and develops integrated HPS. More precisely speaking, I aim at a practical defence of integrated HPS in this book. Rather than providing a full-fledged methodological defence (see Scholl and Räz 2016; McAllister 2018), I would be defending integrated HPS by practising integrated HPS in a concrete historical case study, namely, the case of the origin of genetics. It should be noted that I am not trying to downplay the significance of any methodological defence of integrated HPS, which I do believe is of great importance. However, I contend that a mere methodological defence is not complete or strong enough. An adequate defence should show not only what a good integrated HPS is methodologically, but also how it can be applied in concrete cases. To this end, I argue that the best way to defend integrated HPS is to do it (or practise it). Ultimately, I wish to show how an integrated HPS study helps us to have a better understanding of an historical episode, the origin of genetics, and sheds light on some general issues in the philosophy of science in this book.

Before delving into my integrated HPS work, I would like to say a bit more on my methodology. Although, as I have just emphasised, I am not aiming at a methodological defence, it is necessary to articulate the basics in order to ward off some potential worries. Hasok Chang (2012a, 121–22) proposes two methods of integrated HPS.[6] Method 1 begins with some historiographical puzzle (say, there is a historical episode which is difficult to understand), followed by a search for a new philosophical framework, which provides a better understanding of that episode. Then the new philosophical framework is further developed in order to apply it to other historical episodes. Method 2 begins with a philosophical puzzle, say, a set of putative actions/decisions by past scientists that does not make sense, followed by a search for better historiography, which resolves the philosophical puzzle. Then the new historical account is completed based on empirical work and sheds new light on other related history.

Some chapters in the book will reflect the use of these methods. Nevertheless, there is still something important missing. In Method 1, the integrated HPS analysis begins with a historiographical puzzle, while in Method 2, a search for a better historiography is indispensable. But where are the historiographical puzzles from? And how should we search a better historiography? Typically, both the historiographical puzzles and the better historiographies come from contemporary historical literature. So, shall we start with a review of state of the art of the relevant historical literature? It seems to me not a very satisfactory starting point. Since an integrated HPS work aims at both a good history and a good philosophy, any integrated HPS work should make its own contribution to a good history of science, even for the presentation of a historiographical puzzle. A good integrated HPS work should not assume a particular historical account without argument. Therefore, I argue that, for anyone who plans to use Method 1, he or she should begin writing a historiographical puzzle. Similarly, for anyone who plans to use Method 2, he or she should work on a better historiography.

Surely, there has been a persistent worry concerning philosophers writing history among historians. L. Pearce Williams (1975), for example, famously argues that philosophers should not "be allowed to write history".

> Philosophers tend to be interested in ideas, their logical connections and their logical consequences. They do not seem to find it very interesting to ask where ideas came from, how they developed and how they were interpreted by others who claim to have been influenced by them. They are, therefore, at their best when analysing a system; as we have seen, they are at their worst when trying to account for the evolution of one. (Williams 1975, 252)

Another leading historian of science, I. Bernard Cohen (1977), though having some sympathy to HPS, is concerned with a history written by non-historians, including philosophers.

> Indeed, to many historians, the major danger in the writing of history by nonhistorians (and even by some members of the profession) is the anachronistic application of our present canons of logic and mathematics and of scientific knowledge to prior experiments, laws, and theories. To view the concepts, laws, and theories of a Galileo, a Kepler, and a Newton as

[6]Note that Chang is not only proposing these methods, but also extensively uses these two methods in his integrated HPS study of the chemical revolution (2012b).

'approximations' to some later ideal creations of critical or philosophically minded scientists will block us from a meaningful understanding of the creative processes of any scientist we may be studying, including the interaction of the individual and his social and intellectual environment. (Cohen 1977, 345)

No doubt I oppose Williams' conclusion, but I do think that there is an important lesson from the historians' worry. A good history of science must be holistic and anti-anachronistic. When writing a historical episode, one should not isolate it from its context or understand it in an anachronistic way. Thus, a good way to achieve this is to begin with an anti-anachronistic reading of original texts, correspondences, and relevant published and unpublished writings like a historian would do. Then it is also necessary to examine the relevant secondary literature and scholarly reflections. If a philosopher is writing a good history like a good historian, there is no reason to prevent philosophers from writing history. Therefore, I argue that any serious integrated HPS should begin with an anti-anachronistic reading of history.[7]

There are some obvious benefits of an anti-anachronistic reading of history. For example, it helps to understand the historiographical puzzle in Method 1. In addition, it helps to search for a better historiography in Method 2. And it may also help to inspire a new philosophical framework or a resolution to a philosophical puzzle. It should be noted that an anti-anachronistic reading of history does not guarantee a brand-new historiography. Sometimes it just turns out to confirm the existing historiography. Sometimes it provides a better interpretation in an existing historiographical framework. That said, the significance and the necessity of an anti-anachronistic reading of history should not be overlooked. As I have emphasised, a good integrated HPS work should be a good philosophy of science and a good history of science at the same time. I contend that an anti-anachronistic reading of history provides a good starting point for writing a good history of science as well as doing integrated HPS.

1.3 Integrated HPS in Practice: The Case of the Origin of Genetics

This book will focus on an episode of the history of genetics: the origin of genetics. The origin of genetics is such a mysterious and fascinating topic for both historians of science and philosophers of science, as there are many unsolved puzzles in that period. The traditional narrative of the origin of genetics typically begins with a story of a nineteenth century Austrian monk, Gregor Mendel (1822–1884): Mendel undertook the experiments on *Pisum* (peas) in his garden in Brünn (now Brno, Czech Republic) and discovered the laws of heredity. Mendel presented his discovery at a local scientific society in 1865, and published a paper, entitled *Versuche über Pflanzen-Hybriden* (Experiments on Plant Hybridisation), in a local journal

[7]This is not completely a new idea, but it is often overlooked by philosophers unfortunately.

Verhandlungen Des Naturforschenden Vereins Brünn (Proceedings of the Natural History Society of Brno) in 1866. Unfortunately, Mendel's paper was completely neglected until 1900 when the Dutch biologist Hugo de Vries (1848–1935), the German botanist Carl Correns (1864–1933), and the Austrian agronomist Erich von Tschermak (1871–1962) independently rediscovered it. The rediscovery of Mendel's paper inspired the Cambridge biologist William Bateson (1861–1926) and his associates to develop a Mendelian theory of heredity. However, the Mendelian theory of heredity was immediately resisted by the Biometric School, who advocated a statistical theory of heredity. The leading proponents of Biometry were the UCL statistician Karl Pearson (1857–1936) and the Oxford biologist W. F. R. Weldon (1860–1906). The Mendelian-Biometrician controversy ended soon after the sudden death of Weldon in 1906. Thereafter, the Mendelian theory was developed in great depth and at a rapid speed and eventually incorporated into a highly successful theory of inheritance, classical genetics, by the efforts of the Morgan School in the 1910s and 1920s.

This traditional historiography has been challenged. It has been shown that Mendel's laws were, literally speaking, not about heredity and his concern is no longer simply regarded as an attempt to study the mechanism of heredity (e.g. Olby 1979, 1985; Brannigan 1979; Callender 1988; Monaghan and Corcos 1990). The allegedly rediscovery story is shown to be false. De Vries, Correns, and Tschermak all read Mendel's paper before they wrote their papers (e.g. Rheinberger 1995; Stamhuis et al. 1999; Harwood 2000; Simunek et al. 2011). The significance of Weldon has been recently reassessed (e.g. Radick 2005, 2016; Pence 2011). In short, the historiography of early genetics has been radically revised for the past 50 years. Nevertheless, there is still a lack of consensus on many issues among the historians of science. For example, it is unclear whether Mendel's work could be understood as a study of heredity (e.g. Orel 1996; Olby 1997; Müller-Wille and Orel 2007). It is under debate whether de Vries had discovered the 3:1 ratio independently from his experiments before reading Mendel's paper (e.g. van der Pas 1976; Heimans 1978; Darden 1985; Zevenhuizen 2000). It is puzzling how the history of genetics would have changed if Weldon had not died in 1906 (e.g. Radick 2005). Moreover, the origin of genetics is such a rich historical episode that it may provide many interesting cases for philosophers to examine various issues in general philosophy of science such as conceptual change, confirmation, scientific progress, and theory choice.

In this book, I start my integrated HPS study with a review of the history of genetics from Mendel to Weldon with an anti-anachronistic reading of the original papers, correspondence, and unpublished writings, and a critical examination of the secondary literature. Chapter 2 revisits Mendel's concepts of *Entwicklung* (development) and of *Entwicklungsreihe* (developmental series) in order to re-examine Mendel's work. Chapter 3 examines Mendel's legacy in 1900 and evaluates the significance of de Vries' work in the origin of genetics. Chapter 4 reassesses the historiography of Weldon by focusing on Weldon's late work (1904–1906), especially his unpublished manuscript *Theory of Inheritance*.

Then I move on to two integrated HPS problems: Based on my historical work in Chaps. 2, 3, and 4, what is the best way to analyse and interpret the origin of genetics? What is a best way to understand the progress in the history of genetics if there is any? I call these two problems integrated HPS problems for two reasons. One is that both problems may interest historians of science and philosophers of science. The other is that neither of the problems has been taken seriously by historians of science or philosophers of science today. Many historians of science today are not so interested in the analysis of the pattern of the development of genetics. This sort of work seems a bit too "internal." And philosophers of science are not as interested in developing a philosophical account of the development of science as they were in the 1970s. Nevertheless, I still find both problems worth exploring. A careful analysis of the development and the progress in early genetics will not only provide us a better understanding of the history, but also shed light on the philosophy of science in general. Chapter 5 introduces a new integrated HPS method, the exemplar-based approach, and applies it to analyse and interpret the origin of genetics. Chapter 6 develops a functional approach to the progress in early genetics, motivated by the exemplar-based approach. Chapter 7 offers an exemplar-based explanation of the problem of the long neglect.

Finally, I discuss some general issues in the philosophy of science with my case studies in the history of genetics. Chapter 8 introduces a new mode of conceptual continuity, illustrated by the case of the evolution of the dominance concept. Chapter 9 argues for a practice-based solution to the gap problem in hypothetico-deductivism with the case study of Mendel's evidence for the law of composition of hybrid fertilising cells. Chapter 10 proposes a new criterion of theory choice, illustrated by the case of the Mendelian-Biometrician controversy.

Thus, this book is structured by three parts. Part I consists of three chapters on historical problems. Part II consists of three chapters on integrated HPS problems. Part III consists of three chapters on philosophical problems. Part I provides the historical basis for the discussions in Part II and Part III. Part II offers an integrated HPS method to analyse and interpret the historiography in Part I and to shed new light on the philosophical issues in Part III. Part III develops new philosophical accounts which will in turn make a better sense of the history of sciences more generally.

References

Arabatzis, Theodore. 2017. What's in it for the Historian of Science? Reflections on the Value of Philosophy of Science for History of Science. *International Studies in the Philosophy of Science* 31 (1): 69–82.

Arabatzis, Theodore, and Don Howard. 2015. Introduction: Integrated History and Philosophy of Science in Practice. *Studies in History and Philosophy of Science* 50 (1): 1–3.

Arabatzis, Theodore, and Jutta Schickore. 2012. Introduction: Ways of Integrating History and Philosophy of Science. *Perspectives on Science* 20 (4): 395–408.

Bird, Alexander. 2008. The Historical Turn in the Philosophy of Science. In *The Routledge Companion to Philosophy of Science*, ed. Stathis Psillos and Martin Curd, 67–77. London/ New York: Routledge.
Brannigan, Augustine. 1979. The Reification of Mendel. *Social Studies of Science* 9 (4): 423–454.
Burian, Richard M. 1977. More than a Marriage of Convenience: On the Inextricability of History and Philosophy of Science. *Philosophy of Science* 44 (1): 1–42.
———. 2002. Comments on the Precarious Relationship Between History and Philosophy of Science. *Perspectives on Science* 10 (4): 398–407.
Callender, L.A. 1988. Gregor Mendel: An Opponent of Descent with Modification. *History of Science* 26 (1): 41–75.
Chang, Hasok. 2004. *Inventing Temperature: Measurement and Scientific Progress*. Oxford: Oxford University Press.
———. 2012a. Beyond Case-Studies: History as Philosophy. In *Integrating History and Philosophy of Science: Problems and Prospects*, ed. Seymour Mauskopf and Tad Schmaltz, 109–124. Dordrecht: Springer.
———. 2012b. *Is Water H_2O? Evidence, Realism and Pluralism*. Dordrecht: Springer.
Cohen, I. Bernard. 1977. History and the Philosophy of Science. In *The Structure of Scientific Theories*, edited by Frederick Suppe, 2nd ed., 308–360. Urbana, IL: The University of Illinois Press.
Darden, Lindley. 1985. Hugo de Vries's Lecture Plates and the Discovery of Segregation. *Annals of Science* 42 (3): 233–242.
Friedman, Michael. 2008. History and Philosophy of Sciecne in a New Key. *Isis* 99 (1): 125–134.
Giere, Ronald N. 1973. History and Philosophy of Science: Intimate Relationship or Marriage of Convenience? *The British Journal for the Philosophy of Science* 24 (3): 282–297.
———. 2011. History and Philosophy of Science: Thirty-Five Years Later. In *Integrating History and Philosophy of Science: Problems and Prospects*, ed. Seymour Mauskopf and Tad Schmaltz, 59–65. Dordrecht: Springer.
Grau, Kevin T. 1999. Force and Nature: The Department of History and Philosophy of Science at Indiana University, 1960–1998. *Isis* 90 (S2): S295–S318.
Harwood, Jonathan. 2000. The Rediscovery of Mendelism in Agricultural Context: Erich von Tschermak as Plant-Breeder. *Comptes Rendus de l'Academie des Sciences – Serie III* 323 (12): 1061–1067.
Heimans, Jacob. 1978. Hugo de Vries and the Gene Theory. In *Human Implications of Scientific Advance: Proceedings of the 15th International Congress on the History of Science*, ed. Eric G. Forbes, 469–480. Edinburgh: Edinburgh University Press.
Kuukkanen, Jouni-Matti. 2016. Historicism and the Failure of HPS. *Studies in History and Philosophy of Science* 55: 3–11.
Lakatos, Imre. 1970. History of Science and Its Rational Reconstructions. *PSA: Proceedings of the Biennial Meeting of the Philosophy of Science Association*: 91–136.
Laudan, Larry. 1989. Thoughts on HPS: 20 Years Later. *Studies in History and Philosophy of Science* 20 (1): 9–13.
McAllister, James W. 2018. Using History as Evidence in Philosophy of Science: A Methodological Critique. *Journal of the Philosophy of History* 12 (2): 239–258.
Mendel, Gregor. 1865. Versuche über Pflanzenhybriden. *Verhandlungen des Naturforschenden Vereins Brünn* 4 (Abhandlungen): 3–47.
Monaghan, Floyd V., and Alain F. Corcos. 1990. The Real Objective of Mendel's Paper. *Biology and Philosophy* 5 (3): 267–292.
Müller-Wille, Staffan, and Vitězslav Orel. 2007. From Linnaean Species to Mendelian Factors: Elements of Hybridism, 1751-1870. *Annals of Science* 64 (2): 171–215.
Nickles, Thomas. 1995. Philosophy of Science and History of Science. *Osiris* 10: 138–163.
Olby, Robert Cecil. 1979. Mendel No Mendelian? *History of Science* 17 (1): 53–72.
———. 1985. *Origins of Mendelism*. 2nd ed. Chicago, IL: The University of Chicago Press.

———. 1997. Mendel, Mendelism and Genetics. MendelWeb. *1997*. http://www.mendelweb.org/MWolby.html.

Orel, Vítězslav. 1996. *Gregor Mendel: The First Geneticist*. Oxford: Oxford University Press.

Pas, Peter van der. 1976. Hugo de Vries and Gregor Mendel. *Folia Mendeliana* 11: 229–230.

Pence, Charles H. 2011. 'Describing Our Whole Experience': The Statistical Philosophies of W. F. R. Weldon and Karl Pearson. *Studies in History and Philosophy of Biological and Biomedical Sciences* 42 (4): 475–485.

Pinnick, Cassandra, and George Gale. 2000. Philosophy of Science and History of Science: A Troubling Interaction. *Journal for General Philosophy of Science* 31 (1): 109–125.

Radder, Hans. 1997. Philosophy and History of Science: Beyond the Kuhnian Paradigm. *Studies in History and Philosophy of Science* 28 (4): 633–655.

Radick, Gregory. 2005. Other Histories, Other Biologies. *Royal Institute of Philosophy Supplement* 80 (56): 3–47.

———. 2016. Presidential Address: Experimenting with the Scientific Past. *The British Journal for the History of Science* 49 (2): 153–172.

Rheinberger, Hans-Jörg. 1995. When Did Correns Read Gregor Mendel's Paper? *Isis* 86 (4): 612–616.

Richardson, Alan W. 2008. Scientific Philosophy as a Topic for History of Science. *Isis* 99 (1): 88–96.

Schickore, Jutta. 2009. Studying Justificatory Practice: An Attempt to Integrate the History and Philosophy of Science. *International Studies in the Philosophy of Science* 23 (1): 85–107.

———. 2011. More Thoughts on HPS: Another 20 Years Later. *Perspectives on Science* 19 (4): 453–481.

Scholl, Raphael. 2018. Scenes from a Marriage: On the Confrontation Model of History and Philosophy of Science. *Journal of the Philosophy of History* 12 (2): 212–238.

Scholl, Raphael, and Tim Räz. 2016. Towards a Methodology for Intergrated History and Philosophy of Science. In *The Philosophy of Historical Case Studies*, ed. Tilman Sauer and Raphael Scholl, 69–91. Cham: Springer.

Shapin, Steven, and Simon Schaffer. 2011. *Leviathan and the Air-Pump. Hobbes, Boyle, and the Experimental Life*. Princeton, NJ: Princeton University Press.

Simunek, Michal, Uwe Hoßfeld, and Olaf Breidbach. 2011. 'Rediscovery Revised' – The Coopertation of Erich and Armin von Tschermak-Seysenegg in the Context of the 'Rediscovery' of Mendel's Law in 1899–1901. Plant Biology 13 (6): 835–841.

Stamhuis, Ida H., Onno G. Meijer, and Erik J.A. Zevenhuizen. 1999. Hugo de Vries on Heredity, 1889-1903: Statistics, Mendelian Laws, Pangenes, Mutations. *Isis* 90 (2): 238–267.

Steinle, Friedrich, and Richard M. Burian. 2002. Introduction: History of Science and Philosophy of Science. *Perspectives on Science* 10 (4): 391–397.

Williams, L. Pearce. 1975. Should Philosophers Be Allowed to Write History? *The Bristish Journal for the Philosophy of Science* 26 (3): 241–253.

Wray, K. Brad. 2010. Philosophy of Science: What Are the Key Journals in the Field? *Erkenntnis* 72 (3): 423–430.

Zevenhuizen, Erik. 2000. Keeping and Scrapping: The Story of a Mendelian Lecture Plate of Hugo de Vries. *Annals of Science* 57 (4): 329–352.

Part I
History

We now have a better picture of the origin of genetics than fifty years ago. We know that Mendel's concern was about development (*Entwicklung*) rather than heredity (*Vererbung*). We know that Mendel's paper was not completely neglected before 1900, and de Vries' contribution to the origin of genetics was the introduction of Mendel's work to the study of heredity rather than the rediscovery of Mendel's paper. We know that the development of genetics in the first decade of the twentieth century is more complicated than a debate between Mendelians and Biometricians. However, there are still many puzzles unsolved. It is not clear what development exactly means in Mendel's paper, though we know that Mendel's concern was about development. It is not clear what de Vries' novel contribution is, though we know that de Vries played a vital role in the introduction of Mendel's. It is not clear what Weldon's theory of inheritance is, though we know that the significance of Weldon was unfairly downplayed in the traditional analysis of the Mendelian-Biometrician controversy. In this part, I examine these issues.

Chapter 2
Mendel's *Pisum* Revisited

I decided to return to the original text of Mendel's Versuche über Pflanzen-Hybriden. Here I discovered a new Mendel – a Mendel who did not fit any of the revisionist pictures, but neither was he quite the hero of the traditional account. This was a Mendel who had stepped down from the pages of another story altogether – a different Mendel, who spoke of a different theme. The theme was Entwicklung, which translates into English as development.

(Sandler 2000, 7)

Abstract The historiography of Mendel has changed dramatically for the past five decades. Mendel's paper *Versuche über Pflanzen-Hybriden* is no longer simply viewed as an attempt to study the problem of heredity. It is now a consensus that Mendel's concern, literally speaking, was about the development of hybrids in their progeny. However, there is no consensus on what Mendel meant by development (*Entwicklung*). Nor is there an agreement on the interpretation of Mendel's laws of developmental series (*Entwicklungsreihe*). This chapter revisits Mendel's notions of development and developmental series. Firstly, I argue that Mendel's use of development was greatly influenced by Gärtner's. Secondly, I show Mendel's work on developmental series was novel and important for its new ways of experimentation, conceptualisation, and hypothesisation. Thirdly, I argue that Mendel's laws of developmental series were not about heredity.

Keywords Mendel · Development · Developmental series · Heredity

2.1 Mendel's Concern

To date, Gregor Mendel is still widely credited as the father of genetics and his paper *Versuche über Pflanzen-Hybriden* (Experiments on Plant Hybrids, 1866) is typically viewed as the founding document of the modern study of heredity. However, the historiography of Mendel has changed dramatically for the past five decades. Mendel's

Y. Shan, *Doing Integrated History and Philosophy of Science: A Case Study of the Origin of Genetics*, Boston Studies in the Philosophy and History of Science 320,
https://doi.org/10.1007/978-3-030-50617-9_2

paper is no longer simply viewed as an attempt to study the problem of heredity. It is now a consensus (e.g. Müller-Wille and Orel 2007; Gliboff 2013; Zhang et al. 2017) that Mendel's concern, literally speaking, was about *Entwicklung* (development). It has been widely received (e.g. Olby 1997; Dröscher 2015) that one of Mendel's most important contribution to the history of genetics was his novel work on *Entwicklungsreihe* (developmental series). It has also been well recognised that there are three main sources of the research context of Mendel's work: the Moravian animal breeding community, the Viennese academic botany, and the hybridist tradition (e.g. Gliboff 2013; van Dijk et al. 2018). However, many issues on Mendel and his work still remain puzzling: There is no consensus on what Mendel meant by *Entwicklung* (e.g. Gliboff 1999; Sandler 2000). Nor is there an agreement on the interpretation of Mendel's laws of developmental series (Orel 1996, 1998; Gliboff 1999, 2013; Wood and Orel 2005; Szybalski 2010; van Dijk et al. 2018). This chapter revisits three issues: What was *Entwicklung* meant by Mendel? In what sense was Mendel's conceptualisation of developmental series novel and important? Were Mendel's laws of developmental series about heredity? In Sect. 2.2, I explore the research context of Mendel's notion of *Entwicklung*. In Sect. 2.3, I examine Mendel's notion of developmental series. In Sect. 2.4, I analyse the novelty of Mendel's work on developmental series. In Sect. 2.5, I discuss the interpretations of Mendel's laws of developmental series.

2.2 Mendel and Gärtner on *Entwicklung* (Development)

Mendel explicitly stated his concern in *Einleitende Bemerkungen* (introductory remarks).

> Artificial fertilization undertaken on ornamental plants to obtain new color variants initiated the experiments to be discussed here. The striking regularity with which the same hybrid forms always reappeared whenever fertilization between like species took place suggested further experiments whose task it was to follow the development of hybrids in their progeny.
>
> .... That no generally applicable law of the formation and development of hybrids has yet been successfully formulated can hardly astonish anyone who is acquainted with the extent of the task and who can appreciate the difficulties with which experiments of this kind have to contend. A final decision can be reached only when the results of detailed experiments from the most diverse plant families are available. Whoever surveys the work in this field will come to the conviction that among the numerous experiments not one has been carried out to an extent or in a manner that would make it possible to determine the number of different forms in which hybrid progeny appear, permit classification of these forms in each generation with certainty, and ascertain their numerical interrelationships. It requires a good deal of courage indeed to undertake such a far-reaching task; however, this seems to be the one correct way of finally reaching the solution to a question whose significance for the [developmental][1] history of organic forms must not be underestimated.
>
> The paper discusses the attempt at such a detailed experiment... (Mendel 1866, 3–4, 1966a, 1–2)

[1]In Sherwood's original translation, *Entwicklungs-Geschichte* is translated as "evolutionary history". However, I prefer Staffan Müller-Wille and Kersten Hall's translation as "developmental history" (Mendel 2016), given that the contemporary usage of "evolution" is typically Darwinian-

Table 2.1 Cited Scholars in Mendel's paper (1866)

Cited scholar	The number of occurrences
Kölreuter	6
Gärtner	18
Herbert	1
Lecocq	1
Wichura	3

It seems that the problem of heredity was not Mendel's concern. *Vererbung*, the German word for "inheritance" or "heredity" does not appear in the introductory remarks at all. More surprisingly, it is absent in the rest of the paper, except that Mendel used the verb "*vererbt* (inherited)" once.[2] In contrast, there are two other key words I found. The German word *Hybriden* (hybrids) remarkably appears 101 times. In addition, *Entwicklung* (development) is another key word, appearing 45 times in the paper.[3] These two key words are highly suggestive. They reflect the objective of the paper: to study "the development of hybrids in their progeny". More precisely speaking, Mendel's ultimate aim was to formulate a generally applicable law of the *development of hybrids in their progeny* by a detailed experiment.

The key words also suggest the research context of Mendel's work. As many historians (e.g. Olby 1979, 1985; Brannigan 1979; Müller-Wille and Orel 2007) have already argued, Mendel's work was well within the tradition of hybridism, a research tradition, focusing on the study of plant hybrids by crossing experiments, in the late eighteenth and early nineteenth century.[4] This is also well corroborated by the references that Mendel made in the paper. In the paper, there are only five scholars whose works are mentioned: Joseph Gottlieb Kölreuter, Carl Friedrich von Gärtner, William Herbert, Henri Lecoq, and Max Ernest Wichura (see Table 2.1). Remarkably, all of them were important figures of hybridism. In Mendel's words, they all had "devoted a part of their lives to" the problem of the development of hybrids in their progeny. (Mendel 1866, 3, 1966a, 1–2) What is more, in the concluding remarks (*Schluss-Bemerkungen*), Mendel himself clearly identified that

oriented. As I shall show later, Mendel's *Entwicklung* was nothing to do with Darwin's theory of evolution.

[2]The original German text is "auch beschränkt sich diese Eigentümlichkeit nur auf das Individuum und vererbt sich nicht auf die Nachkommen". (Mendel 1866, 14) (Eva R. Sherwood's translation (1966a, 12): "furthermore, this peculiarity is restricted to the individual and not inherited by the offspring.")

[3]It should be noted that *Entwicklung* is also a key word in Mendel's correspondence to Nägeli, appearing 19 times, either as an independent noun or as an element of a compound word.

[4]The central problem of hybridism is what distinguishes species from accidental varieties. The leading figures of hybridism include Carl Linnaeus (1707–1778), Joseph Gottlieb Kölreuter (1733–1806), and Carl Friedrich von Gärtner (1772–1850). For a detailed study of the history of hybridism, see Roberts (1929), Olby (1985), and Müller-Wille and Orel (2007).

his work on *Pisum* (i.e. peas) was within the "field" of the hybridist tradition, led by "two authorities" Kölreuter and Gärtner,[5] and made a lengthy comparison of his work with theirs.

Although it is now a received view that Mendel's concern was about the development of hybrids in their progeny, there is no agreement on the interpretation of Mendel's concern. Given its significance, I find it necessary to make clear the meaning of Mendel's *Entwicklung* at first.

In Mendel's paper, *Entwicklung* appears 22 times as an independent noun, and 23 times as an element in a compound word (e.g. *Entwicklungsgeschichte*, *Entwicklungsreihe*, and *Entwicklungs-Gesetz*). As a noun, it is usually translated as development, though occasionally as formation (Mendel 1966a, 33, 43) or constitution (Mendel 1965, 19, 38). Remarkably, *Entwicklungsgeschichte* is not accordingly translated as developmental history or history of development. Rather it is typically translated as evolutionary history (Mendel 1966a, 2, 41) or history of evolution (Mendel 1965, 35).[6] Such a translation leads to a once popular but mistaken reading of Mendel's work and its context. For example, based on the English translations of the same passage in which *Entwicklungsgeschichte* is translated as evolutionary history, Margaret Campbell (1982, 40) and L. A. Callender (1988, 51) take for granted that Mendel's work should be understood within a Darwinian evolutionary context. Such a reading is too arbitrary, however. It is problematic to conflate the nineteenth century German word *Entwicklungsgeschichte* with the nineteenth century English word evolution without argument. As Robert Olby (1997) points out, "it is very misleading to transpose Mendel's work from its source in the Austro-Hungarian empire to the world of Darwinian debates in Victorian England and America." Thus, in order to figure out the very meaning of Mendel's *Entwicklung*, it seems necessary to locate it in its intellectual context.

Sander Gliboff (1999) proposes that Mendel's use of *Entwicklung* was directly influenced by Franz Unger.[7] For Gliboff, the connection between Unger and Mendel is both intellectual and sociological. It is recorded that Unger taught Mendel botany at University of Vienna in the period 1851–1853. Unger is also thought to be one of a few people to whom Mendel sent an offprint of his paper.[8] Both of Unger and Mendel were connected to some academic associations (e.g. the Society of Naturalists in Brno and the Zoological-Botanical Society in Vienna) and involved in some academic activities (e.g. a project of surveying the sprawling Habsburg Empire).

---

[5]"A comparison of the observations made on *Pisum* with the experimental results obtained by Kölreuter and Gärtner, the two authorities in this field, cannot fail to be of interest." (Mendel 1966a, 39)

[6]Müller-Wille and Hall's translation (Mendel 2016) corrects this.

[7]Unger (1800–1870) was an Austrian botanist, notable for his pre-Darwinian theory of evolution (1852b).

[8]Gliboff (1999, 234f33) admits that there is some doubt about this, though.

Thus, Gliboff (1999, 226) argues that under the influence of Unger, Mendel referred *Entwicklung* to both "the individual ontogeny and the evolution of lineage."[9]

A glimpse of Unger's work seems to be compatible with Gliboff's conclusion. *Entwicklung* is also a key word in Unger's work, appearing 21 times in *Botanische Briefe* (Botanical Letters, Unger 1852a) and in 47 times in *Versuch einer Geschichte der Pflanzenwelt* (Experiment of a History of the Plant World, Unger 1852b). More surprisingly, I find that the phrases *Entwicklungsgeschichte* (developmental history),[10] *Entwicklungsgesetze* (developmental laws), and *Entwicklungsreihe* (developmental series) were already used by Unger.[11] Thus, it seems that Gliboff's argument that Mendel's use of *Entwicklung* was Unger-oriented is plausible, given the connection between Unger and Mendel. However, another puzzle arises. If Mendel's concern was directly influenced by Unger, why was Unger not cited or mentioned at all in Mendel's paper?

A more careful reading of Unger's work (1852a, b) and Mendel's paper (1866) suggests something more complicated. It is clear that Unger and Mendel used the terms *Entwicklung* (development), *Entwicklungsgeschichte* (developmental history), *Entwicklungsgesetze* (developmental laws), and *Entwicklungsreihe* (developmental series) differently. Although both of Unger and Mendel used *Entwicklung* (development) to designate individual ontogeny in some cases, Unger talked much of it at the cellular level (e.g. Unger 1852a, 104, 106, 112) while Mendel at the morphological level (Mendel 1866, 8, 11). Unger (1852a, 110) referred *Entwicklungsreihe* (developmental series) to a series of developmental process, while Mendel (1866, 5, 20, 21, 22, 24, 29, 31, 35, 39, 40) referred *Entwicklungsreihe* (developmental series) to the statistical proportion of hybrid forms. Unger's *Entwicklungsgesetze* (developmental laws) was about the whole plant world (Unger 1852b, 282), while Mendel's was specifically about hybrids (Mendel 1866, 18, 32). Mendel (1866, 4) was explicit on the point that his work on *Pisum* was significant for "*die Entwicklungs-Geschichte der organischen Formen* (the developmental history of organic forms)", but it is too hasty to conclude that this was related to Unger's general interest in "*die Entwicklungsgeschichte der Pflanzenwelt* (the developmental history of plant world)", especially given that Mendel was implicit on in what sense his work would shed new light on "*die Entwicklungs-Geschichte der organischen Formen* (the developmental history of organic forms)." What is more, the other key word *Hybriden* is completely missing in Unger's work (1852a, b). Therefore, it is dubious that Mendel's concern on hybrid development or his use of *Entwicklung* was directly influenced by Unger.

---

[9] Individual ontogeny is the development of an organism, usually a process from the fertilisation of the egg to the organism's mature form.

[10] This shows that Gliboff's claim (1999, 235f42) that the term "*Entwicklungsreihe*" seems to be Mendel's own coinage is problematic.

[11] The phrase *Entwicklungsgeschichte* appears nine times, *Entwicklungsgesetze* once, and *Entwicklungsreihe* once in Unger (1852a).

In contrast, Mendel did explicitly relate his use of *Entwicklung* to Gärtner's both in his paper and in the correspondence to Carl Wilhelm Nägeli.

> Gärtner mentions that in cases where development was regular the two parental types themselves were not represented among the offspring of the hybrids, only occasional individual closely approximating them. (Mendel 1866, 40, 1966a, 40–41)
>
> The results which Gärtner obtained in his experiments are known to me; I have repeated his work and have re-examined it carefully to find, if possible, an agreement with those laws of development which I found to be true for my experimental plant. (Mendel 1966b, 57)

These passages clearly show that Mendel shared the use of *Entwicklung* with Gärtner (at least in some cases). It seems not a big surprise, given that Gärtner is the most cited scholar (18 times) in Mendel's paper. However, it is a bit surprising that nobody has yet attempted to study the meaning of *Entwickelung*[12] in Gärtner's book and its influence on Mendel's use. Thus, I contend that it is worth studying Gärtner's use of *Entewickelung* in his book carefully for the purpose of making clear Mendel's use of *Entewicklung*.

Entwickelung really is one of the central terms in Gärtner's book *Versuche und Beobachtungen über die Bastarderzeugung im Pflanzenreich* (Experiments and Observations on the Hybridisation in the Plant World, 1849). The root word *Entwicke* appears 332 times in the book. In Gärtner's book, *Entwickelung* is definitely nothing to do with evolution (whether in a Darwinian sense or an Ungerian sense[13]). Rather, it is closer to what we now refer to as individual ontogeny. In most cases, Gärtner designated *Entwickelung* to a process of the growth of the plant, or of a specific part of the plant (e.g. ovary, embryo, and flower). Here are examples.[14]

> In contrast, in the case of natural fertilisation, although all parts of the female organs have not yet reached their full development, the pollination of the stigma with their own pollen has rarely been unsuccessful.[15] (Gärtner 1849, 9)
>
> If the interior of a hybrid fruit is examined in the first period of its development, the fertilised seeds are not found in the same degree of development and size.[16] (Gärtner 1849, 29)

---

[12] *Entwickelung* is an old spelling of *Entwicklung*.

[13] When talking of the Ungerian evolution, I have Unger's concept of *Entwicklungsgeschichte* (developmental history) in mind. Unger (1852a, b) spoke much of *Entwicklungsgeschichte* (developmental history), which, as Gliboff (1999, 226) correctly points out, referred to changes in the flora through the geological time.

[14] Gärtner's book (1849) is not yet translated into English. If not indicated otherwise, all the translations of Gärtner's text are mine.

[15] "..., da im Gegentheil bei der natürlichen Befruchtung, wenn auch alle Theile der weiblichen Organe ihre vollstandige Entwickelung noch nicht erlangt haben, eine Bestäubung der Narbe mit dem eigenen Pollen sehr selten erfolglos bleibt, ..."

[16] "Wenn eine solche durch Bastardzeugung entstandene Frucht in der ersten Periode ihrer Entwickelung im Innern untersucht wird, so findet man die befruchteten Eichen nicht in gleichem Grade der Entwickelung und der Grösse."

> These experiments seem to show once again that in addition to the various invisible developmental states of the female organs of plants, both of the sunlight and the heat . . . have a great influence on the course of the fertilisation of plants.[17] (Gärtner 1849, 49)
>
> This doubt arises, especially in the case of hybrids: Do the defective pollens possess the power to affect the development of the outer envelopes of the fruit and the seed?[18] (Gärtner 1849, 98)
>
> For the four plants of this kind, which had grown from the same seed and the same pod, all the flower-heads were at the same time castrated before their development and maturity of the anthers occurred at the same time.[19] (Gärtner 1849, 566)

It is clear that Mendel used the term *Entwicklung* in a similar way. For example,

> A defective development of the keel has also been observed. (Mendel 1866, 5, 1966a, 8)
>
> In the pods first formed by a small number of plants only a few seeds developed, . . . (Mendel 1866, 13, 1966a, 11)

In addition to *Entwicklung*, *Hybriden* (hybrids), the other key word in Mendel's paper, is also a central term in Gärtner's book, in which *Hybriden* appears 176 times and its synonym *Bastarden* appears 362 times. The overlap of the key words indicates a common interest: Both of Gärtner's book and Mendel's paper were about hybrids and their development, as the titles suggest.[20] Both Gärtner and Mendel talked much of laws of hybrid development, though they used the phrases slightly different. Mendel consistently spoke of *Entwicklungsgesetz* (developmental law), while Gärtner used the phrases *Entwicklungsgesetz* (developmental law) and *Gesetz der Formbildung der Bastarde* (law of the formation of hybrids) interchangeably.[21] What is more, Mendel's view on the law of hybrid development was very similar to Gärtner's. For instance, Gärtner strongly believed that the formation and development of hybrids are based on certain laws (*die Entwickelung und Bildung einer jeden Pflanze beruhe auf gewissen Gesetzen*), while those laws were still not yet known.

> The general laws of development of the growth of the parts in hybridisation do not seem to change; all the changes in the hybrid plant-bodies follow the same laws as in the pure species.[22] (Gärtner 1849, 363)

---

[17]"Diese Versuche scheinen abermals zu zeigen, dass neben den verschiedenen, dem Auge unsichtbaren Entwickelungsgraden der weiblichen Organe der Gewächse, die beide Agentien, das Sonnenlicht und die Wärme, (s. oben S. 10) einen grossen Einfluss auf den Gang der Befruchtung der Pflanzen haben."

[18]"Hier tritt namentlich bei den Hybriden der Zweifel ein: ob nicht auch der taube Pollen die Kraft besitze, die Entwickelung der äusseren Umhüllungen der Frucht und der Samen zu bewirken."

[19]"An vier Pflanzen dieser Art, welche aus dem gleichen Samen aus einer und derselben Schote aufgegangen waren, wurden alle Blumenknöpfe vor ihrer Entwickelung und eingetretenen Reife der Antheren zu gleicher Zeit castrirt."

[20]Recall the title of Mendel's paper is "Experiments on Plant Hybrids", while the title of Gärtner's book is "Experiments and Observations on Hybrid Formation in the Plant Kingdom."

[21]Gärtner sometimes used the phrases *Entwickelung* (development) and *Bildung* (formation) interchangeably (e.g. Gärtner 1849, 585).

[22]"Die allgemeinen Entwickelungsgesetze der Theile der Gewächse scheinen daher durch diehybride Zeugung keine, den Sinnen perceptible Aenderung zu erfahren; sondern alle

Given original relation of plant and environment, which determines the complete development of the species, is lost, the deviation of a plant from its normal type is the necessary consequence of *the development and formation of each plant which are based on certain laws*, and these laws, necessary for the perfect development of a plant, are expressed in the different proportions of the external factors, light, moisture, soil, air quality, heat, etc. Yet we certainly do not know these laws; but their existence is by no means questioned, especially since they are confirmed rather by a variety of phenomena.[23] (Gärtner 1849, 494)[24]

Rather, we hope and believe that with the help of hybridisation we will find and discover the laws of development of plants. . .[25] (Gärtner 1849, 605)

This view was also reflected by Mendel in his introductory remarks, and strengthened in several places later.

No generally applicable *law of the formation and development of hybrids* has yet been successfully formulated. (Mendel 1866, 3, 1966a, 2)

Anyone surveying the shades of color that appear in ornamental plants as a result of like fertilization cannot easily escape the conviction that . . . development proceeds according to [certain laws].[26] (Mendel 1866, 38, 1966a, 38)

. . . unity in the plan of development of organic life is beyond doubt.[27] (Mendel 1866, 43, 1966a, 43)

Mendel's conviction that a search for the law of hybrid development[28] was important for the study of "the developmental history of organic forms" seems to echo Gärtner's view that the laws of hybrid development were helpful to solve the problems of species-forms and of hybrid-forms.

Since we still lack the means to explain the development of the various plant forms from the simple cell to the perfect *development of the various forms of plants* in their various phases to follow or construct them in the organism, we are not yet able to do so to determine the

Entwickelungen und Veranderungen des hybriden Pflanzenkorpers nach denselben Gesetzen zu erfolgen, wie bei den reinen Arten."

[23]"Werde dieses ursprüngliche, die vollständige Entwickelung, ja die Existenz der Art bedingende, Verhältniss aufgehoben, so sei die Abweichung einer Pflanze von ihrem Normaltypus die nothwendige Folge davon, d. i. die Entwickelung und Bildung einer jeden Pflanze beruhe auf gewissen Gesetzen, und werde durch diese bedingt, und diese Gesetze sprechen sich aus in den, zur vollkommenen Entwickelung einer Pflanze nothigen, verschiedenen Verhaltnissen der Einwirkung der ausseren Momente, Licht, Feuchtigkeit, Boden, Luftbeschaffenheit, Wärme u. s. w. Noch kennen wir freilich diese Gesetze so gut als gar nicht; ihr Vorhandensein lasse sich aber durchaus nicht mehr verkennen, wir seien vielmehr durch eine Menge von Erscheinungen gezwungen, sie als vorhanden anzunehmen."

[24]This is a quote from Hornschuh, but it is clear that Gärtner shared this view.

[25]"wir hoffen und glauben vielmehr, dass wir mit Hulfe der Bastardzeugung zur Auffindung und Entdeckung der Formgesetze der Gewächse gelangen werden."

[26]Sherwood's original translation is that "according to a certain law", but it is in fact a mistranslation, because in Mendel's German text the plural form *Gesetze* (laws) is used.

[27]". . . die Einheit im Entwicklungsplane des organischen Lebens ausser Frage steht."

[28]Mendel never explicitly defined *Gesetz* (law), but he contended that a law should be something universally applicable and studied empirically.

correlation, with which the mechanism of hybrid development is related to the vegetable transmutation in general.[29] (Gärtner 1849, 293)

Doesn't the continuity and reality of a system of plants depend on the stability throughout generations? ... If the plant-species are something transitory and changeable, their development of forms is not something solid, grounded in nature, but is so much dependent on external influences that the basic form of one species changes in the course of time and may change into a completely different form.

It seems to us that this vital question of systematic botany can be decided upon from the vegetation itself and from the laws of development of plants without having to wait for an answer in a millennium.[30] (Gärtner 1849, 605)

Moreover, the objective of Mendel's paper as searching for the law of the development of hybrids in their progeny seems to follow a question asked by Gärtner at the end of the book.

How do these different seeds *behave in their further development* (in 1849) with respect to the type of plants and their seed production?[31] (Gärtner 1849, 680)

Considering the similarity of the uses of *Entwick(e)lung* and the views on the law of hybrid development, and the textual connections between Mendel's and Gärtner's work, I argue that Mendel's usage of *Entwicklung* had been inherited from (or at least greatly influenced by) Gärtner's. In particular, as I have shown, both Mendel's and Gärtner's *Entwick(e)lung* were about hybrids rather than about the plant world as a whole. In other words, I argue that Mendel's usage of *Entwicklung* was Gärtnerian-oriented rather than Ungerian-oriented.

It is worth noting that although I argue that Mendel's use of *Entwicklung* was inherited from Gärtner's rather than Unger's, I am not trying to dismiss Unger's influence on Mendel. I am sympathetic to the view that Mendel's work on *Pisum* was to some extent influenced by Unger. For example, as Gliboff (1999) and Ariane Dröscher (2015) show, Mendel's mathematical approach was indebted to Unger's quantitative approach to botany. However, I do not think that we should

[29] "Da es uns noch an Mitteln fehlt, die Entstehung und Entwickelung der verschiedenen Pflanzenformen von der einfachen Zelle an bis zur vollendeten Entwickelung des vollkommenen Gewächses in ihren verschiedenen Phasen zu erklären und im Organismus zu verfolgen oder zu construiren: so sind wir auch noch nicht im Stande, die Bande zu bestimmen, womit der Metaschematismus der hybriden Bildung mit der vegetabilischen Metamorphose überhaupt zusammenhängt."

[30] "Beruht dann nicht die Dauer und Wirklichkeit eines Systems der Gewächse auf der Stabilitat in der Art von Generation zu Generation? Würde das Streben und die Arbeit der Systematiker aller Zeiten und die kostbaren Iconographien nicht zur blosen Spielerei herabsinken und völlig unnutz sein? wenn die Pflanzenart etwas Vergangliches und Wandelbares, ihre Gestaltsbildung nicht etwas Festes, in der innersten Natur Begründetes, sondern von äusseren Einwirkungen soweit Abhangiges ware, dass die Grundform einer Art im Laufe der Zeiten sich andern, in eine ganz andere Gestalt übergehen, und in ein ganz anderes Wesen sich verwandeln würde.

Es scheint uns, dass diese Lebensfrage der systematischen Botanik aus der Vegetation selbst und aus den Gesetzen der Formbildung der Gewächse werde entschieden werden konnen, ohne auf die Entscheidung von Jahrtausenden warten zu mussen."

[31] "Wie sich diese verschiedenen Samen in ihrer weiteren Entwickelung (im Jahr 1849) in Absicht auf den Typus der Pflanzen und ihrer Samenerzeugung verhalten werden."

overestimate Unger's influence on Mendel's work. My reading of Mendel's, Gärtner's, and Unger's work show that Mendel's use of *Entwicklung* was much closer to Gärtner's than to Unger's. In short, there is little textual evidence to support Gliboff's reading that Mendel's use of *Entwicklung* was influenced by Unger.[32] Therefore, contra Gliboff, I argue that Mendel's use of *Entwicklung* was inherited from Gärtner's rather than Unger's.

2.3 Mendel's "*Entwicklungsreihe* (Developmental Series)"

Although I argue that Mendel's use of *Entwicklung* was to a great extent influenced by Gärtner's, it does not imply that Mendel's concern (1866) was identical with Gärtner's (1849). Nor was Mendel's work simply a continuation of Gärtner's. Gärtner's main concern in his 1849 book was the problem of the distinction between species and accidental varieties, a central problem of hybridism. The problem originated from Linnaeus's short essay *Plantae hybridae* (1751), which is regarded as "the founding document of the hybridist tradition" (Müller-Wille and Orel 2007, 177). However, unlike his hybridist predecessors (e.g. Linnaeus 1751; Kölreuter 1763), Gärtner adopted a new approach. According to him,

> The question of what distinguishes species from varieties is therefore [. . .] a purely biological one: a secure foundation for determining species cannot be found solely in abstraction, neither in the characters, nor in the intermediate forms, but has to be sought in reflection, that is in the individual history (*individuellen Geschichte*) of each species, its whole development (*Entwickelung*), and not in a particular aspect only. (Gärtner 1849, 151; Müller-Wille and Orel 2007, 187)

Note that this was the first time in history that the problem of the species/varieties distinction was studied by examining "the development of various forms of plants (*die Entwickelung der verschiedenen Pflanzenformen*)". (Gärtner 1849) Thus, it provides another piece of evidence that Mendel's use of *Entwicklung* was influenced by Gärtner's.

However, there is a crucial difference between Gärtner's and Mendel's concerns. Gärtner focused on the development of hybrids in one generation, while Mendel was particularly interested in the patterns of the development of hybrids in the successive generations. It should be highlighted that Mendel particularly referred "the development of hybrids in their progeny" to "the developmental series" (*Entwicklungsreihe*) of hybrid forms in the following generations (i.e. the statistical distribution of different morphological forms). Mendel noted that the law of development of hybrid in their progeny could be only be discovered by determining the

[32]For a similar reason, I find Iris Sandler's claim (2000, 9) that Mendel's use of *Entwicklung* was influenced by M. J. Schleiden untenable. Sandler's reason is that "as a botanist [Mendel] would have been familiar with the textbook written by the leading botanist of the period, M. J. Schleiden… Its influence was widespread." To me, such a speculation is too bold.

"numerical relationships of different forms of hybrids". He also explicitly mentioned that the numerical relationships of hybrid forms were determined by observing the developmental series of offspring.

> To discover the relationships of hybrid forms to each other and to their parental types it seems necessary to observe without exception all members of the series[33] (*Entwicklungsreihe*) of offspring in each generation. (Mendel 1866, 5, 1966a, 4)

Thus, Mendel's concern can also be summarised as a study of the developmental series of hybrid in successive generations, where the development series means the proportion of the different forms of hybrids. Accordingly, a crucial difference between Gärtner's and Mendel's work can be roughly summarised as that Gärtner took a qualitative approach to hybrid development, while Mendel a quantitative one.

It is clear that Mendel's major discussions in the paper were centred on the developmental series.

> If A denotes one of the two constant traits, for example, the dominating one, a the recessive, the Aa the hybrid form in which both are united, then the expression
>
> A + 2Aa +a
>
> gives the [developmental series] for the progeny of plants hybrid in a pair of differing traits. (Mendel 1866, 17, 1966a, 16)
>
> When, therefore, two kinds of differing traits are combined in hybrids, the progeny develop according to the expression:
>
> AB + Ab + aB + ab + 2ABb + 2aBb + 2 AaB + 2Aab + 4AaBb
>
> Indisputably this [developmental series] is a combination series in which the two [developmental series] for the traits A and a, B and b are combined term by term. (Mendel 1866, 20–21, 1966a, 20)
>
> The difference of forms among the progeny of hybrids, as well as the ratios in which they are observed, find an adequate explanation in the principle just deduced. The simplest case is given by the [developmental series] for one pair of differing traits. (Mendel 1866, 29, 1966a, 29)

Moreover, all Mendel's laws were in fact about the developmental series. In his paper, Mendel formulated three laws of "development of hybrid" explicitly: the law of development (*Entwicklungs-Gesetz*) that "apply to a pair of differing traits (*welche nur in einem wesentlichen Merkmale verschieden waren*)" (Mendel 1866, 18), the "law of combination of differing traits (*Gesetz der Combinirung der differierenden Merkmale*)" (Mendel 1866, 32), and the law of "the composition of hybrid fertilizing cells (*die Beschaffenheit der hybriden-Befruchtungszellen*)" (Mendel 1866, 45). The law of development concerning a pair of differing traits (LDT) was formulated as follows:

---

[33]This is a major flaw in Sherwood's translation (Mendel 1966a), in which *Entwicklungsreihe* is translated as series rather than developmental series in all of its 17 occurrences. Clearly, such a translation fails to reflect the significance of *Entwicklungsreihe* (or even *Entwicklung*) in Mendel's paper (1866).

> ... [O]f the seeds formed by the hybrids with one pair of differing traits, one half again develop the hybrid form while the other half yield plants that remain constant and receive the dominating and the recessive character in equal shares. (Mendel 1866, 17, 1966a, 15)

The law of combination of differing traits (LCT) was stated as follows.

> The progeny of hybrids in which several essentially different traits are united represent the terms of a combination series in which the [developmental series] for each pair of differing are combined... at the same time that the behavior of each pair of differing traits in a hybrid association is independent of all other differences in the two parental plants. (Mendel 1866, 22, 1966a, 22)

The law of composition of hybrid fertilising cells (LCC) was formulated as follows.

> ... [P]ea hybrids form germinal and pollen cells that in their composition correspond in equal numbers to all the constant forms resulting from the combination of traits united through fertilization. (Mendel 1866, 29, 1966a, 29)

It is clear that Mendel's laws were all about the developmental series. For example, LDT was about the developmental series of dominant constant, hybrid, and recessive constant forms of hybrids, which was also symbolically formulated by Mendel as A + 2Aa + a where A denoted the dominant constant form, Aa the hybrid form, and a the recessive constant form.[34] Mendel's LCT was about the developmental series in the progeny of hybrids which differs more than a pair of differing traits, while Mendel's LCC provided a reductive explanation of LCT and LCD. Thus, it seems to be more appropriate to call Mendel's laws "the laws of developmental series". This, again, confirms my argument that Mendel's concern was about developmental series in the progeny of hybrids. Therefore, I argue that when Mendel talked of the development of hybrids in their progeny, he was referring to the developmental series in the progeny of hybrids.[35]

This is also why Mendel explicitly identified that his task is "to follow the development of hybrids in their progeny" rather than "to follow the development of hybrids themselves" in the introductory remarks. That is, as I shall elaborate in more detail in the next section, Mendel was taking a quantitative approach to the problem of hybrid development. Thus, Mendel's objective as looking for the law of the developmental series was definitely a creative extension of Gärtner's research. As Staffan Müller-Wille and Vitězslav Orel (2007, 192) put it, "Mendel, in stating his aims, was simply taking the programme of Gärtner a step forward." His real concern followed and developed Gärtner's interest in the development of the plant, where "the developmental history of organic forms" had been explicitly defined by Gärtner as a process from the single cell to a perfectly mature form of a plant (Gärtner 1849, 293).

---

[34]For an elaboration of this, see Sect. 2.4.

[35]Unfortunately, partly because of the traditional mistranslation of *Entwicklungsreihe* (Bateson 1902; Mendel 1966a), historians used to overlook the relation of "developmental series (*Entwicklungsreihe*)" and "the development of hybrids in their progeny (*die Entwicklung der Hybriden in ihren Nachkommen*)".

2.4 Mendel's Novel Conceptualisation: The Laws of Developmental Series

Some historians argue that most of Mendel's work was nothing astonishingly new. Most of his work on *Pisum* was merely a confirmation of observations reported before.

> Before Mendel, the component parts of Mendelism had been discovered separately, some by the plant hybridizers and some by the bee breeders. (Zirkle 1951, 103)
>
> [Mendel's] observations on segregation and independent assortment were recorded by his predecessors and the focus on inheritance ratios was pioneered by his contemporary. (Brannigan 1979, 440)

However, this is definitely not what Mendel himself thought of his work on *Pisum*. In a letter to Nägeli (dated 18 April 1867), Mendel was clear on the point that he believed that he did discover something novel or revolutionary.

> I knew that the results I obtained were not easily compatible with our contemporary scientific knowledge. (Correns 1906, 199; Mendel 1966b, 60)

Based on his carefully designed experiments, Mendel noticed the "striking regularity" of the development of hybrids in their progeny from his experiments on *Pisum*. For example, Mendel recognised that the hybrid seeds of purely bred yellow peas and green ones were all yellow. What is more, Mendel conceptualised the observation that all the hybrids were yellow. He denoted that yellowness in the parental peas as a dominant parental trait, which referred to the parental trait passing unchanged to all of the offspring, while greenness as a recessive parental trait, which referred to the parental trait absent in the hybrid offspring.

Moreover, when these hybrid seeds were self-fertilised, both yellow and green seeds were obtained in the offspring. And the ratio of the yellow seeds to the green ones was close to 3: 1 (see Table 2.2). Accordingly, Mendel proposed that the ratio of the seeds with the dominant trait to the ones with the recessive trait was 3: 1. It must be also emphasised that it was not trivial for Mendel to recognise those Mendelian ratios. As we can see from Table 2.2, though all the ratios were close

Table 2.2 The Result of the First Generation from the Hybrid Acquired by Mendel (1866)

Experiment	Amount (of seeds with one trait)	Amount (of seeds with the other trait)	Ratio
1	5474 (round)	1850 (wrinkled)	2.96: 1
2	6022 (yellow)	2001 (green)	3.01: 1
3	705 (grey-brown)	224 (white)	3.15: 1
4	882 (inflated)	299 (constricted)	2.95: 1
5	428 (green)	152 (yellow)	2.82: 1
6	651 (axial)	207 (terminal)	3.14: 1
7	787 (long)	277 (short)	2.84: 1

to 3: 1, it was still a novel move for Mendel to make such a statistical inference.[36] In addition, Mendel's recognition of the 3: 1 ratio was more than a simple approximation or idealisation of the raw data. Rather it was a conceptual analysis in terms of dominance and recessiveness. Without the definition of dominant and recessive traits, the 3: 1 ratio was unrecognisable. It was a substantial conceptual construction by Mendel to classify the morphological traits in terms of dominance and recessiveness. What is more, Mendel further reconceptualised the 3: 1 ratio into the 1: 2: 1 ratio, which represented the distribution of dominant (parental), dominant (hybrid), and recessive (parental) traits.

> [T]he average ratio between the number of forms with the dominating trait and those with the recessive one is . . . 3: 1.
>
> The dominating trait can have double significance here – namely that of the parental characteristic or that of the hybrid trait. In which of the two meanings it appears in each individual case only the following generation can decide. As parental trait it would pass unchanged to all of the offspring; as hybrid trait, on the other hand, it would exhibit the same behavior as it did in the first generation. (Mendel 1866, 14–15, 1966a, 13)
>
> The ratio of 3: 1 in which the distribution of the dominating and recessive traits take place in the first generation therefore resolves itself into the ratio of 2: 1: 1 in all experiments if one differentiates between the meaning of the dominating trait as a hybrid trait and as a parental trait. (Mendel 1866, 16–17, 1966a, 15)

In these paragraphs, the concept of dominance was creatively redefined. Mendel distinguishes two senses of the dominant trait. The dominant parental trait (or the dominant form) was the trait which passed unchanged to all of the offspring, while the dominant hybrid trait (or the hybrid form) which would exhibit the same behaviour with the 3: 1 ratio in its offspring, as illustrated in Fig. 2.1, where A

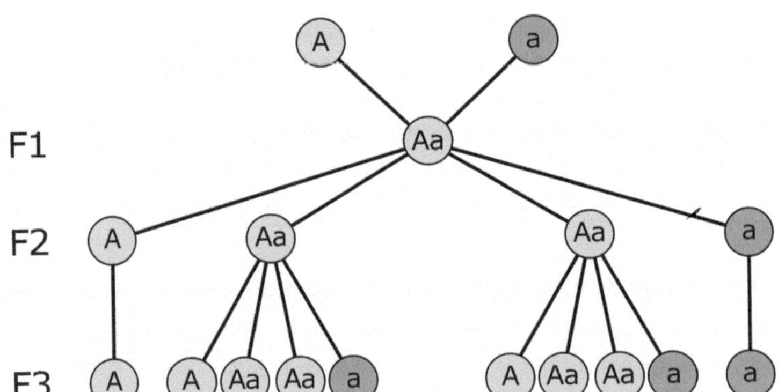

Fig. 2.1 Dominance in Pea Hybridisation

[36]Three decades later, Hugo de Vries, when undertaking the similar crossing experiments, initially failed to recognise the 3: 1 ratio. Based on the results of his crossing experiments on *Lychnis vespertina* glabra × *Lynchnis diurnal* in 1894, de Vries (1897, 72) claimed that the ratio of the hairy seedlings and hairless seedling is 2: 1. However, 3 years later, de Vries (1900b, 75) modified it as a 3: 1 ratio.

denoted the dominant parental trait, while Aa the dominant hybrid trait. (In contemporary terminology, Mendel referred A, a, and Aa to different morphological forms.)

This redefinition was really important for Mendel. It suggests that he recognised that there was a distinction between the yellow seeds in the F1 generation. Some yellow seeds only produced yellow seeds, while others produced both yellow and green seeds. The former was redefined as the dominant parental trait, whereas the latter as the dominant hybrid trait. This distinction led Mendel to recognise another "striking" regularity. Among the offspring of the peas with the dominant hybrid trait, the distribution of the dominant parental trait, dominant hybrid trait, and the recessive trait was 1: 2: 1 again. Based on this, Mendel formulated LDT, which he wished to be applicable universally. Thus, I argue that Mendel's work on developmental series was novel in three significant ways. Firstly, Mendel designed and undertook a series of experiments, which produced massive data for his study of developmental series. Though seeking information through experimentation was not something new in hybridism, Mendel introduced a novel way of experimenting hybrids in order to study the developmental series, which was influential among the early Mendelians such as Carl Correns and Erich von Tschermak.[37] Secondly, Mendel made a novel use of the concept of dominance to classify the data obtained from the experiments.[38] Thirdly, Mendel used a new mathematical approach[39] to studying the developmental series, and proposed the laws of developmental series. In short, as Lenny Moss (2003, 23) summarises, "Mendel's paper illustrates an exemplar for how to set up an empirical practice."

Now it is the time to highlight the significance of the Mendelian ratios. Though the phenomenon of dominance had been observed by many (e.g. Knight 1799; Goss 1824; Seton 1824) since the late eighteenth century, Mendel was the first to conceptualise the phenomenon in terms of dominance/recessiveness, and record, analyse, and explain the statistical relation of dominant, hybrid, and recessive traits. Such a statistical analysis of the dominant and recessive traits was introduced into the study of heredity around 1900, especially by de Vries (1900a, c, d) and William Bateson (1902), which preluded the birth of genetics.

---

[37]For example, Mendel's experimental design that planting peas both in the field and int pots indoors to protect plants from foreign contamination was adopted by Tschermak (see von Tschermak 1900, 233).

[38]For a detailed discussion, see Chap. 8.

[39]Note that by arguing that Mendel's mathematical approach is novel, I do not mean to argue that Mendel was the first to use the mathematical or statistical approach to biological study. As Dröscher (2015) shows, mathematical thinking was not as alien as thought in the nineteenth century biology. I argue that Mendel was creative for introducing the mathematical approach to the problem of hybrid development in two senses. Firstly, Mendel's mathematical approach was different from his contemporaries' like Unger's or Nägeli's. Secondly, Mendel is the first to use a mathematical approach to hybrid development. Although both Gärtner and Mendel focused on hybrid development, Gärtner was taking a qualitative approach, while Mendel a quantitative one.

2.5 Mendel and the Study of Heredity

From a contemporary point of view, it seems that Mendel's laws of developmental series were obviously about transmission of hereditable traits. Thus, some historians (e.g. Orel 2009; Orel and Peaslee 2015; van Dijk et al. 2018) maintain that Mendel's laws were in fact about heredity. As Sandler (2000, 11) argues, "Mendel's intention – 'to follow the development of hybrids in their progeny' – a step-by-step description of the transmission and distribution of hybrid traits between parent and progeny. Is it not fitting that we restore to Mendel his well-deserved title – Father of Genetics?" Such an interpretation of Mendel's work on developmental series is fairly popular, because it seems to fit perfectly with the historiography of Mendel as a founder of genetics.

There are two main lines of argument for that Mendel's laws were in fact about heredity. One is presented by Raphael Falk and Sahotra Sarkar (1991), who, though accepting that there are substantially conceptual differences between Mendel's laws and the Mendelian laws of inheritance,[40] argue that Mendel was studying the problem of heredity in terms of development.

> Indeed, as Olby ... has observed, Mendel phrased his problem in terms of the formulation of hybrids and their progeny. The reason for this is the historical context: in the first half of the nineteenth century, Moravia was a center of intensive breeding activity which provoked considerable interest in intellectual circles ... The breeding methods of Robert [Bakewell] that were imported from England and promoted by Geisslern (known as the "Moravian [Bakewell]") were those of the production of hybrids between divergent strains showing desired traits and transmit them to the progeny over several generations. A difficulty that arose was that the traits did not breed true. When Mendel addressed such problems he was, therefore, directly addressing a problem of heredity. Conceptually, moreover, it could not have been otherwise. If hybrids are formed through reproduction, and pass traits on (with whatever success) through reproduction, and these are the traits being studied, what is being studied, *ipso facto*, is the inheritance of traits. The problem of inheritance is, in some sense, more general than the problem of hybridization. But that hardly means that studying hybridization is not studying inheritance. (Falk and Sarkar 1991, 448)

At first glance, Falk and Sarkar's argument seems to be promising and interesting. Unfortunately, it is seriously flawed. The very problem is its anachronistic premise. Falk and Sarkar are looking at Mendel's problem with a twentieth century lens. In other words, Mendel's problem is situated in a present-day understanding of the problem of inheritance. Falk and Sarkar's argument can be reformulated as follows:

P1. Transmission of morphological traits is a central problem in the science of heredity.

---

[40]Mendel's laws are what were articulated in Mendel's paper, while the Mendelian laws of inheritance are what were developed by the early Mendelians (e.g. Bateson) and classical geneticists (e.g. T. H. Morgan). There are some obvious differences between Mendel's laws and the Mendelian laws. For example, the Mendelian laws were typically formulated in terms of genes, while Mendel did not have a concept of gene. For a detailed discussion, see Olby (1979, 1985).

P2. Mendel was studying transmission of the morphological trait of *Pisum* in terms of development.

C. Therefore, Mendel's work was (or at least can be understood as) a study of heredity.

However, I have to warn that our current understanding of the problem of inheritance is heavily influenced by Mendel's approach. The problem of inheritance is indebted to Mendel's work. Under the influence of Mendel's focus on the transmission of morphological traits, geneticists in the early twentieth century began taking transmission as a central research problem in the study of heredity, which consists in our current understanding of the science of heredity. Therefore, it is anachronistic (and even circular) to argue that Mendel's work was about heredity by arguing that Mendel's problem is similar to the problem of transmission inheritance today!

Another line of argument for that Mendel's laws of developmental series were about heredity runs as follows: In the first half of the nineteenth century there were lively discussions on heredity in Brünn, Moravia, where Mendel was undertaking his research. The interest of heredity arose from the study of sheep breeding. The term *genetische Gesetze* (genetic laws) was first coined in 1818 to describe the patterns of inheritance in animals by Count E. Festetics. Since 1827, the word *Vererbung* (heredity) had been widely used to describe the transmission of different traits. Johann Karl Nestler (1783–1841), Professor of Agriculture, Science and Natural History at the Moravian University of Olomouc, Franz Diebl (1770–1859), Professor of the Philosophical Institute, and Franz Cyrill Napp (1792–1867), abbot of the Augustinian monastery in Brünn, were three key figures in the study of heredity at that time. It is argued that Mendel must have been familiar with the context, given that Napp was Mendel's superior, and Mendel attended Diebl's lectures. Therefore, Mendel's work on developmental series was about heredity and he "reformulated and tried to answer Napp's question 'What is inherited and how?'" (Orel and Wood 2000, 1041).

Orel, a strong proponent of this view, reinforces this view by arguing that a key term in Mendel's paper *Entwicklungsgeschichte* (developmental history) should be identical with *Verebungsgeschichte* (history of heredity).

> At that time prominent sheep breeders in Moravia had kept forty years of stock registers with wool sample cards. Nestler called on them to take part in the elaboration of the principles of rational breeding to answer the key question: "What noticeable success in heredity can be achieved when rams and ewes with equal or unequal traits are paired?" The breeders were asked to examine these old family registers to investigate the history of heredity (*Verebungsgeschichte*) of the best stock animals in their offspring from the top downward or their developmental history (*Entwicklungsgeschichte*) in their ancestors from bottom upward. From this investigation Nestler expected valuable material for the theory of breeding. The term *Entwicklungsgeschichte* was for him the other side of the same coin, of *Verebungsgeschichte*. (Orel 1998, p. 297)

> Emphasising the significance of his research approach from the view point of "Entwicklungsgeschichte of organic forms", Mendel could have had in mind Nestler's understanding of the history of heredity. (Orel 1998, p. 299)

I agree with Orel that the problem of heredity was important in the context of animal breeding in Moravia. As Roger Wood and Orel point out,

> The big problem facing [breeders in Moravia], . . . was the absence of a theory of inheritance. In 1836 Napp stated his opinion that the problem could be explained only by seeking its physiological basis, i.e. by discovering the nature and behaviour of whatever it was that was transmitted at fertilisation. When discussion on this topic continued in the following year, he formulated the key research question 'what is inherited and how?' (Wood and Orel 2005, 268)

It is true that there were some discussions on *Vererbung* (heredity) in the literature of animal breeding in Moravia in the first half of the nineteenth century (e.g. Nestler 1837). It is also plausible to postulate that Mendel might have known the work on *Vererbung* (heredity) by Nestler and Napp, through either his personal acquaintance with Napp or his study under Nestler. Nevertheless, it is still unknown to what extent Mendel was influenced by these studies on heredity: Did Mendel regard the problem of heredity as an interesting problem to study? The Sheep Breeder's Association in Brünn (SBA) was the main forum for the discussion on the problem of heredity in the first half of the nineteenth century. However, the sudden death of Nestler, a leader of SBA, in 1841, marks the end of activities in animal breeding somehow. As a result, the discussion on heredity in Moravia diminished. As Orel (1977, 195) admits, there was only occasional publication on heredity after 1840. Thus, it is very doubtful that Mendel's concern on developmental series was trying to revive an interest of heredity which was dead for at least a decade.

Given the textual evidence we have so far, it is too bold to infer that Mendel's research on the development of plant hybrids was a means to studying the problem of heredity under the influence of the Morvarian sheep breeders. No direct evidence shows that Mendel's paper was related to *the problem of heredity* studied by Nestler and Napp. Otherwise, why didn't Mendel mention their works in the paper? Why didn't Mendel even suggest the potential contribution made by his laws of developmental series to the problem of heredity? Why didn't Mendel make a comparison between his observation on peas and the work of Nestler or other breeders in the concluding remarks, as he did with Kölreuter and Gärtner?

Peter J. van Dijk, Franz J. Weissing, and T. H. Noel Ellis seem to have the answers.

> Mendel's experiments had more implications, which Mendel discussed in his paper, such as the transformation of one species into another, the cytology of fertilization, the generation of variation by the conditions of life vs. hybridization, speciation, and the stability of species and hybrids. All these reflect also Mendel's interest in pure science. According to the report of the second *Pisum* lecture in the *Mährischer Korrespondent*, Mendel first gave an introduction to (what was known about) "the cell and the reproduction of the plants by fertilization." before he presented his own research (Anonymous 1865a, b). Therefore, it makes sense that Mendel chose the broad title "Experiments on Plant Hybrids," without specifically mentioning heredity or inheritance. Therefore, it makes sense that Mendel chose the broad title "Experiments on Plant Hybrids," without specifically mentioning heredity or inheritance. Mendel's broad interest in plant biology was clearly sanctioned by Napp's comments relating to the need for a scientific study of inheritance. (van Dijk et al. 2018, 353)

Furthermore, a sentence in a letter written by Nägeli to Mendel is quoted as a piece of evidence that Mendel's work was about heredity.

> Although the word inheritance was used only once in the text of the Pisum paper and was missing from the title, the paper is unmistakably about the rules of inheritance. That was quite clear to Nägeli when he wrote to Mendel: "I am convinced that with many forms you will get notably different results (in respect to the inherited characters)." (van Dijk et al. 2018, 353)

It is correct that Mendel's paper came across so many different topics. It is also correct that Mendel's interests were broad. However, it should be noted that not only the word *Vererbung* was missing, but also no discussions on heredity was found in Mendel's paper. It is definitely not "clear" that Mendel's laws were about heredity.

To sum up, it can be concluded that Mendel's work on developmental series was not about heredity. For those who have not yet been convinced by my arguments and are still inclined to maintain that Mendel's laws were about heredity, I would like to highlight the puzzles again.

If Mendel's real concern would have been about heredity, why didn't Mendel emphasise this literally in the paper or in his correspondence with Nägeli? Why didn't Mendel's contemporaries, especially those who were interested in the problem of heredity, read the paper as a study of heredity?[41] It seems to me really difficult to insist that Mendel's work was in fact about heredity until these puzzles are well solved.[42]

2.6 Conclusion

As many historians (e.g. Olby 1979; Orel and Wood 1998; Gliboff 1999; Dröscher 2015) have successfully shown, Mendel's work was not from nowhere. Mendel's work was greatly influenced by his predecessors' and contemporaries' work. So was Mendel's use of the phrase *Entwicklung*. In a nutshell, I have argued that Mendel's use of *Entwicklung* was greatly influenced by Gärtner's, in which *Entwicklung* is defined as a process from the single cell to a perfectly mature form of a plant (Gärtner 1849, 293). I also argued that Mendel's real concern was to determine the developmental series of the forms of hybrids in the progeny. And I have shown that Mendel's work on developmental series was novel for its new ways of experimentation, conceptualisation, and hypothesisation. Finally, I argued that Mendel's laws of developmental series were not about heredity, despite its great influence on the history of genetics.

---

[41]Recent studies on the reports of Mendel's lecture in 1865 by local newspaper (Zhang et al. 2017; van Dijk et al. 2018) confirm my scepticism. None of the five articles read Mendel's work as a study of heredity.

[42]It should be noted that I am not trying to dismiss the significance of Mendel's work in the history of genetics. I am happy with the historiography of Mendel as a founder of genetics, but resist the interpretation that Mendel was interested in heredity or Mendel's work was about heredity. For my interpretation, see Chap. 5.

References

Anonymous. 1865a. Monats-Versammlung des Naturforschenden Vereins. Mährischer Correspondent 5 (57): 4 (10 March 1865). http://www.digitalniknihovna.cz/mzk/view/uuid:8d8b4bc0-64c8-11e3-8c6a-005056825209?page=uuid:3a0ad420-6711-11e3-8387-001018b5eb5c

———. 1865b. Monats-Versammlung des Naturforschenden Vereins in Brünn am 8. März 1865. Brünner Zeitung 65, 362 (20. March 1865). http://www.digitalniknihovna.cz/mzk/view/uuid:e32cb9e0-f061-11e3-a012-005056825209?page=uuid:59ba8440-f7f9-11e3-8232-5ef3fc9ae867

Bateson, William. 1902. *Mendel's Principles of Heredity: A Defence*. Cambridge: Cambridge University Press.

Brannigan, Augustine. 1979. The Reification of Mendel. *Social Studies of Science* 9 (4): 423–454.

Callender, L.A. 1988. Gregor Mendel: An Opponent of Descent with Modification. *History of Science* 26 (1): 41–75.

Campbell, Margaret. 1982. Mendel's Theory: Its Context and Plausibility. *Centaurus* 26 (1): 38–69.

Correns, Carl. 1906. Gregor Mendels Briefe an Carl Nägeli, 1866–1873. *Abhandlungen der Mathematisch-Physischen Klasse der Königlich Sächsischen Gesellschaft der Wissenschaften* 29 (3): 189–265.

de Vries, Hugo. 1897. Erfelijke Monstrositeiten in den Ruilhandel der Botanische Tuinen. *Botanisch Jaarboek* 4: 62–93.

———. 1900a. Das Spaltungsgesetz der Bastarde (Vorlaufige Mittheilung). *Berichte der Deutschen Botanischen Gesellschaft* 18 (3): 83–90.

———. 1900b. Hybridising of Monstroities. *Journal of the Royal Horticultural Society* 24: 69–75.

———. 1900c. Sur la Loi de Disjonction des Hybrides. *Comptes Rendus de l'Academie des Sciences (Paris)* 130: 845–847.

———. 1900d. Sur les Unités des Caractères Spécifiques et Leur Application à l'étude des Hybrides. *Revue Génerate de Botanique* 12: 257–271.

Dröscher, Adriane. 2015. Gregor Mendel, Franz Unger, Carl Nägeli and the Magic of Numbers. *History of Science* 53 (4): 492–508.

Falk, Raphael, and Sahotra Sarkar. 1991. The Real Objective of Mendel's Paper: A Response to Monaghan and Corcos. *Biology and Philosophy* 6 (4): 447–451.

Gärtner, Carl Friedrich. 1849. *Versuche und Beobachtungen über die Bastarderzeugung im Pflanzenreich*. Stuttgart: Gedruckt bei K. F. Hering & Comp.

Gliboff, Sander. 1999. Gregor Mendel and the Laws of Evolution. *History of Science* 37 (2): 217–235.

———. 2013. The Many Sides of Gregor Mendel. In *Outsider Scientists: Routes to Innovation in Biology*, ed. Oren Harman and Michael R. Dietrich, 27–44. Chicago and London: The University of Chicago Press.

Goss, John. 1824. On the Variation in the Colour of Peas, Occasioned by Cross Impregnation. *Transactions of the Horticultural Society of London* 5: 234–236.

Knight, Thomas Andrew. 1799. An Account of Some Experiments on the Fecundation of Vegetables. *Philosocial Transactions of the Royal Society of London* 89: 195–204.

Kölreuter, Joseph Gottlieb. 1763. *Fortsetzung der Vorläufigen Nachricht von Einigen das Geschlecht der Pflanzen betreffenden Versuchen und Beobachtungen*. Trans. Staffan Müller-Wille and Vitězslav Orel. Leipzig.

Linnaeus, Carl. 1751. Plantae Hybridae. In Caroli Linnaei Ammoenitates Academicae, Seu Dissertationes Variae Physicae, *Medicae, Botanicae Antehac Seorsim Editae*. Trans. Staffan Müller-Wille and Vitězslav Orel, 3rd ed., 28–62. Stockholm.

Mendel, Gregor. 1866. Versuche über Pflanzenhybriden. *Verhandlungen des Naturforschenden Vereins Brünn* IV (1865) (Abhandlungen): 3–47.

———. 1965. Experiments in Plant Hybridisation. Trans. Royal Horticultural Society of London. Cambridge, MA: Harvard University Press.

———. 1966a. Experiments on Plant Hybrids. In *The Origin of Genetics: A Mendel Source Book*, edited by Curt Stern, Eva R. Sherwood, translated by Eva R. Sherwood, 1–48. San Francisco, CA: W. H. Freeman and Company.

———. 1966b. Gregor Mendel's Letters to Carl Nageli: 1866–1873. In *The Origin of Genetics: A Mendel Source Book*, edited by Curt Stern, Eva R. Sherwood, translated by Leonie Kellen Piternick and George Piternick, 56–102. San Francisco, CA: W. H. Freeman and Company.

———. 2016. Experiments on Plant Hybrids (1866). Translated by Staffan Müller-Wille and Kersten Hall. British Society for the History of Science Translation Series. 2016. https://www.bshs.org.uk/bshs-translations/mendel/.

Moss, Lenny. 2003. *What Genes Can't Do*. Cambridge, MA: The MIT Press.

Müller-Wille, Staffan, and Vitězslav Orel. 2007. From Linnaean Species to Mendelian Factors: Elements of Hybridism, 1751-1870. *Annals of Science* 64 (2): 171–215.

Nestler, J.K. 1837. Ueber Vererbung in der Schafzucht. *Mittheilungen der Gesllschaft Zur Beförderung des Ackerbaues, der Natur- und Landeskunde in Brünn* 34, 35, 36: 265–69, 273–79, 281–86, 289–93, 300–303, 318–3.

Olby, Robert Cecil. 1979. Mendel No Mendelian? *History of Science* 17 (1): 53–72.

———. 1985. *Origins of Mendelism*. 2nd ed. Chicago, IL: The University of Chicago Press.

———. 1997. "Mendel, Mendelism and Genetics." *MendelWeb. 1997.* http://www.mendelweb.org/MWolby.html.

Orel, Vitězslav. 1977. Selection Practice and Theory of Heredity in Moravia Before Mendel. *Folia Mendeliana* 12: 179–200.

———. 1996. *Gregor Mendel: The First Geneticist*. Oxford: Oxford University Press.

———. 1998. Constant Hybrids in Mendel's Research. *History and Philosophy of the Life Sciences* 20 (3): 291–299.

———. 2009. The 'Useful Questions of Heredity' Before Mendel. *Journal of Heredity* 100 (4): 421–423.

Orel, Vitězslav, and Margaret H. Peaslee. 2015. Mendel's Research Legacy in the Broader Historical Network. *Science & Education* 24 (1–2): 9–27.

Orel, Vitězslav, and Roger J. Wood. 1998. Empirical Genetical Laws Published in Brno Before Mendel Was Born. *Journal of Heredity* 89 (1): 79–82.

———. 2000. Essence and Origin of Mendel's Discovery. *Comptes Rendus de l'Academie des Sciences – Serie III* 323 (12): 1037–1041.

Roberts, H. F. 1929. *Plant Hybridization Before Mendel*. Princeton, NJ: Princeton University Press.

Sandler, Iris. 2000. Mendel's Legacy to Genetics. *Genetics* 154 (1): 7–11.

Seton, A. 1824. Note by the Secretary. *Transactions of the Horticultural Society of London* 5: 236–237.

Szybalski, W. 2010. Professor Alexander Zawadzki of Lvov University – Gregor Mendel' s Mentor and Inspirer. *Biopolymers and Cell* 26 (2): 83–86.

Unger, Franz. 1852a. *Botanische Briefe*. Vienna: Verlag von Carl Gerold & Sohn.

———. 1852b. *Versuch einer Geschichte der Pflanzenwelt*. Vienna: Wilhelm Braumüller.

van Dijk, Peter J., Franz J. Weissing, and T.H. Noel Ellis. 2018. How Mendel 's Interest in Inheritance Grew out of Plant Improvement. *Genetics* 210 (October): 347–355.

von Tschermak, Erich. 1900. Über Künstliche Kreuzung bei Pisum Sativum. *Berichte der Deutschen Botanischen Gesellschaft* 18 (6): 232–239.

Wood, Roger J., and Vitězslav Orel. 2005. Scientific Breeding in Central Europe During the Early Nineteenth Century: Background to Mendel's Later Work. *Journal of the History of Biology* 38 (2): 239–272.

Zhang, Hui, Wen Chen, and Kun Sun. 2017. Mendelism: New Insights from Gregor Mendel's Lectures in Brno. *Genetics* 207 (September): 1–8.

Zirkle, Conway. 1951. Gregor Mendel & His Precursors. *Isis* 42 (2): 97–104.

Chapter 3
De Vries' Mendelism Reassessed

It is important to realize that De Vries was in a unique position compared to other researchers in the early history of genetics, because he was the only one who had worked out a detailed theory of heredity before the rediscovery of Mendel's laws... De Vries did not really come to terms with Mendel, because he did not distance himself from his own theory. He reacted in several manners. He tried to bring the content of the two approaches into line with each other. He attempted to do this in two different ways but in both cases he was not successful.

(Stamhuis 2015, 47)

Abstract It is now received that de Vries' work is better characterised as the incorporation of Mendel's work into his project of the theory of pangenesis. However, there is still lack of a systematic evaluation of de Vries' incorporation. It is unclear in what sense de Vries' incorporation was significant for the origin of genetics. Nor is it very clear what role Mendel's work played in de Vries' incorporation. Would there have been a Mendelian theory of heredity in the first decade of the twentieth century if Mendel had not ever written or published his paper on *Pisum*? This chapter examines the significance of de Vries' work and its relation to Mendel's work. Firstly, I argue that Mendel's work was indispensable for de Vries' conceptualisation and theorising. Secondly, I show that de Vries' unsuccessful incorporation was still important for its introduction of Mendel's work to the study of heredity and the proposal of the law of segregation.

Keywords De Vries · Segregation · Mendel · Mendelism · Dominance · Activeness

3.1 The Rediscovery Story

The famous story that Gregor Mendel's work on *Pisum* was hardly noticed until 1900 when Hugo de Vries, Carl Correns, and Erich von Tschermak independently re-discovered it was originally kept by all three's advocacy in their papers.

Y. Shan, *Doing Integrated History and Philosophy of Science: A Case Study of the Origin of Genetics*, Boston Studies in the Philosophy and History of Science 320, https://doi.org/10.1007/978-3-030-50617-9_3

> This important treatise [i.e. Mendel's paper (1866)] is so seldom cited, that I myself for the first time came to know about it after I had closed the majority of my experiments, and had derived therefrom the principles contributed in the text. (de Vries 1900a, 85 n1, 1966, 110 n6)
>
> When I discovered the regularity of the phenomena, and the explanation thereof... the same thing happened to me which now seems to be happening to de Vries: I thought that I had found something new. But then I convinced myself that the Abbot Gregor Mendel in Brünn, had, during the sixties, not only obtained the same result through extensive experiments with peas, which lasted for many years, as did de Vries and I, but had also given exactly the same explanation, as far as that was possible in 1866. (Correns 1900, 158, 1966, 119–20)
>
> Correns has just published experiments, which also deal with artificial hybridization of different varieties of *Pisum sativum* and observations of the hybrids left to self-fertilization through several generations. They confirm, just as my own, Mendel's teachings. The simultaneously "discovery" of Mendel by Correns, de Vries, and myself appears to me especially gratifying. Even in the second year of experimentation, I too still believed that I had found something new. (Tschermak 1900a, 239, 1950, 47)

However, all these claims have been undermined by historians. It has been shown that Mendel's work was not completely neglected in the nineteenth century academia. There were at least a dozen references to Mendel's paper before 1900 (e.g. Olby 1985, 219–34; Weiling 1991, 10–11; Orel 1996, 275–79). It has also been shown that none of de Vries', Correns', and Tschemark's work can be legitimately regarded as an "independent" rediscovery of Mendel's work. All of de Vries', Correns', and Tschemark had already read Mendel's paper before the completion of (at least the majority of) their research (e.g. Monaghan and Corcos 1986; Rheinberger 1995; Zevenhuizen 2000). It has been argued that "rediscovery" is not an appropriate term to describe or summarise de Vries', Correns', and Tschemark's work in 1900 (e.g. Meijer 1985; Bowler 1989; Olby 1989; Stamhuis et al. 1999). As Robert Olby (1989, 208) points out, "[The concept] of rediscovery tells us so little about the history of the establishment of Mendelian genetics that we should cease to organize our histories of genetics around it." Then, what happened in 1900? How should this episode be best understood?

Recently, the rediscovery story has been reshaped as three parallelled but intertwined stories of incorporation. De Vries, Correns, and Tschermak had different research interests and backgrounds, so they read and treated Mendel differently. Correns' paper (1900) was clearly an attempt to test Mendel's approach and theory. As William Bateson (1902, 14) well summarised, Correns' paper was a report of a repetition of "Mendel's original experiment with Peas having seeds of different colours." Tschermak's papers (1900a, b) are shown to be the attempts to incorporate the Mendelian approach into his project of a theory of regular differential valency of traits in heredity with the help of his brother Armin von Tschermak (Simunek et al. 2011, 2012; Gliboff 2015). It has also been suggested that de Vries' work is better characterised as the incorporation of Mendel's work into his project of the theory of pangenesis (e.g. Theunissen 1994; Stamhuis 2015). However, there is still lack of a systematic evaluation of de Vries' incorporation. It is unclear in what sense de Vries' incorporation was significant for the origin of genetics. Nor is it very clear what role Mendel's work played in de Vries' incorporation. Would there have been a

Mendelian theory of heredity in the first decade of the twentieth century if Mendel had not ever written or published his paper on *Pisum*? This chapter reexamines the significance of de Vries' work and its relation to Mendel's work. In Sect. 3.2, I argue that Mendel's work was indispensable for de Vries' theorising of the law of segregation. In Sect. 3.3, I show that de Vries' incorporation, though unsuccessful, was important for its introduction of the Mendelian approach to the study of heredity and the proposal of the law of segregation.

3.2 No Mendel, No Mendelians!

De Vries (1900a, 90) was explicit on that what he rediscovered are the following two statements.

> 1. Of the two antagonistic characteristics, the hybrid carries only one, and that in complete development. Thus in this respect the hybrid is indistinguishable from one of the two parents, There are no transitional forms.
> 2. In the formation of pollen and ovules the two antagonistic characteristics separate, following the most part simple laws of probability. (de Vries 1900a, 84, 1966, 110)

He claimed that both were drawn by Mendel and deduced the law of segregation from claim 2.

> In the formation of pollen grains and ovules these characters separate. The individual pairs of antagonistic characters behave independently during this process. From this separation the law can de deduced:
>
> *The pollen grains and ovules of monohybrids are not hybrids* but belong exclusively to one or the other of the two parental types. (de Vries 1900a, 86, 1966, 112)

However, de Vries' claim has been shown to be problematic in various ways. Firstly, Mendel never made a similar statement to claim 2. Moreover, none of Mendel's three laws can be simply identified with an early version of claim 2. As shown in Chap. 2, Mendel's laws were either about the statistical distribution of the morphological traits of hybrids and their progeny, or about the correspondence between the combination of traits and the composition of cells. In contrast, de Vries' claim 2 was literally about the formation of the pollen grains and ovules in the generative period. In short, Mendel did not explicitly state out the process of segregation as de Vries elaborated in claim 2. Nor did Mendel ever use the term segregation (*Spaltung*). As I shall argue in Sect. 3.3.3, it was de Vries' contribution to coin the term "law of segregation (*Spaltungsgesetz*)" and to designate it to a phenomenon in the formation of pollen grains and ovules. In other words, de Vries' claim that he rediscovered Mendel's law of segregation was literally false. Secondly, it has been shown that de Vries' reading of Mendel's paper substantially influenced his formation of the law of segregation (e.g. Kottler 1979; Zevenhuizen 2000). Thirdly, it has been shown that de Vries' law of segregation is better understood as an attempt to incorporate Mendel's approach into his work on pangenesis (e.g. Meijer 1985; Theunissen 1994; Stamhuis et al. 1999; Stamhuis 2015).

That said, it seems to some that a minimal sense of de Vries's rediscovery claim is still defensible. It can be argued that de Vries had discovered the Mendelian ratios[1] independently.

3.2.1 The 3: 1 Ratio in the 1896 Notes

Ida Stamhuis, Onno Meijer, and Erik Zevenhuizen find some notes dated 1896 in the Hugo de Vries Archive, which are claimed to contain the evidence of de Vries' recognition of the 3: 1 ratio and his proposal of the law of segregation. In one of the sheets containing information on *Aster tripolium*, dated August 10, 1896, de Vries wrote:

> Discussion. According to the law of pangene hybridization (p. 187), the purple specimens from white mother must have purple fathers and be central hybrids. They show therefore that the white specimens at Huizen (preferably almost entirely, partially? [inserted: 95%]) have been fertilized by purple ones. Just as my Trifol. pat. alb. have been fertilized by my 7-leaved race at 16b VI.
>
> Therefore gain seed and sow it. If there are no white flowers this year and therefore all specimens are central hybrids, the seed must give 75% purples and 25% whites. This to be investigated.
>
> At the same time this is a new principle in the transfer of varieties from the wild into the garden. If this happens after fertilization in the field, then all specimens from the seed can be look like old-types; then the variant will still emerge from their seed (namely in 25% of the specimens).[2] (Stamhuiset al. 1999, 249–50)

In another note, dealing with *Veronica longifolia*, dated August 18, 1896, de Vries wrote:

> According to the 1.2.1 law the old types, no matter how they are pollinated, always give have to give 100% blue ones, while the central bastards in the case of free pollination (with the exception of whites), therefore by central bastards and by old-types, would have to give between 0 and 25% whites. (Stamhuis et al. 1999, 250)

At first glance, this discovery seems to be well consistent with de Vries' own claim that he first obtained the 3: 1 ratio and the explanation in 1896.

> This law [i.e. the law of segregation of hybrids] is not new. It was stated, for a particular case (peas), more than thirty years ago. It is Gregor Mendel who formulated it in a paper entitled "Versuche uber Pflanzen-Hybrids" in the journal *Verhandlungen des naturforschenden Vereins in Brünn (T IV, p. 1), 1865*. Mendel made the derivations not only for monohybrid, but also for di-and poly-hybrids.
>
> This memoir, too good for its time, was overlooked and forgotten. We find only rare citations, and then only for auxiliary observations. I also did not have knowledge of it until after the completion of the most essential part of the experiments discussed in this article,

[1] As discussed in Chap. 2, one of Mendel's achievement was his conceptualisation of the 3: 1 ratio in terms of dominance and recessivness.

[2] The English translation quoted is from Stamhuis, Meijer and Zevenhuizen's paper (1999).

> and particularly after having found the demonstration of principle from the fourth year of cultivation (1896) of my poppy crosses.
>
> I hope, however, that I have shown that Mendel's law not only applies to peas, but also is applicable in a very general way to all true hybrids.[3] (de Vries 1900d, 271)

Moreover, both "the law of pangene hybridization" and "the 1.2.1 law" look so similar to the law of segregation, and easily make one to suspect that they were the premature versions of the law of segregation. Based on these two notes, Stamhuiset al. (1999) conclude that in 1896 de Vries already discovered the 3: 1 ratio and the Mendelian explanation.

If this analysis is correct, the significance of Mendel's work for the origin of genetics should be downplayed. As Lindley Darden (1976, 167) famously speculates, "if Mendel had not existed, the laws very likely would have been called 'De Vries's Laws' and history would have proceeded as it did, *sans* references to Mendel."

I firmly oppose this counterfactual conclusion. I argue that Mendel's work was indispensable for de Vries' proposal of the law of segregation. Had de Vries never read Mendel's paper, there would be no law of segregation published in 1900. I argue that de Vries' 1896 notes do not suffice to show de Vries' independence of the discovery of the Mendelian ratios or of the law of segregation. What de Vries's 1896 notes show is merely that he might have recognised that there are some 3: 1 ratios in his hybridisation experiments. However, it is worth noting that these 3: 1 ratios are not identical with the Mendelian ratios in de Vries' papers (1900a, c, d). Before delving into the discussion, I find it necessary to clarify three different things: the 3: 1 ratio in de Vries' 1896 notes, Mendel's 3: 1 ratio (1866, 14–15), and de Vries' Mendelian 3: 1 ratio in his 1900 papers (e.g. 1900a, 86–87).

It is clear that the 3: 1 ratio in the 1896 notes was substantially different from Mendel's 3: 1 ratio and de Vries' Mendelian 3: 1 ratio. Both of Mendel's 3: 1 ratio and de Vries' Mendelian 3: 1 ratio were conceptualised in terms of dominance and recessiveness.

> When the results of all experiments are summarized, the average ratio between the number of forms with the dominating trait and those with the recessive one is 2.98: 1, or 3: 1. (Mendel 1866, 14–15, 1966, 13)

[3]This is my translation of de Vries' text: "Cette loi n'est pas nouvelle. Elle a été énoncée, pour un cas particulier (les pois), il y a plus de trente années. C'est Gregor Mendel qui l'a formulé dans un mémoire intitule « Versuche über Pflanzen-Hybriden », inséré dans le journal Verhandlungen d. nat. Vereins in Brünn (T IV, p. 1), 1865. Mendel en a déduit les conséquences non seulement pour les monohybrides, mais aussi pour les di-polyhybrides.

Ce mémoire, trop beau pour son temps, a été méconnu et oublié. On ne le trouve cité que rarement, et alors seulement pour des observations accessoires. Aussi n'en ai-je pris connaissance moi-mème qu'après avoir achevé la partie la plus essentielle des expériences citées dans cet article, et notamment, après avoir trouvé la démonstration du principe par la quatrième année de culture (1896) de mes pavots croisés.

J'espère cependant avoir démontré que la loi de Mendel ne vaut pas seulement pour les pois, mais qu'elle s'applique d'une manière très générale à tous les vrais hybrides." (de Vries 1900d, 271)

> According to our first main statement the hybrids possess the dominanting trait, so that one obtains 75% of individuals with the dominating trait, 25% of individuals with the recessive trait. (de Vries 1900a, 87, 1966, 113).

These ratios were more than a statistical summary. Rather they were theoretical constructs. And they were part of the broader theoretical frameworks respectively. As I have shown in Sect. 2.4, Mendel's 3: 1 ratio was further analysable into the 1: 2: 1 ratio, which was a mathematical representation of Mendel's law of combination of differing traits (LDT), while de Vries' Mendelian 3: 1 ratio was in fact a derivation from the law of segregation. However, the 3: 1 ratio in the 1896 notes was merely descriptive. It is implicit how it was related to de Vries' general theoretical framework in a broader context, say, the theory of pangenesis (1889). Nor is it clear how the 3: 1 ratio in de Vries' 1896 notes was related to the law of segregation, especially given that, as Stamhuis et al. (1999, 251) point out, in 1896 de Vries did not have the concepts of dominance and of recessiveness.

In addition, the law of pangene hybridization and the 1.2.1 law mentioned in the notes are dubious. These two laws are found nowhere except the 1896 notes. It is still unclear what these laws exactly meant. Stamhuis et al. (1999, 251–52) suggest that when de Vries talked of the 1.2.1 law, he had Adolphe Quetelet's binomial law (1871) in mind. Even if this is true, it is still unjustified to identify the Queteletan-oriented 3: 1 ratio with the Mendelian 3: 1 ratio derived from the law of segregation. Moreover, it is worth noting that there was no 3: 1 ratio explicitly mentioned in the 1896 notes. What de Vries in fact wrote was "the seed must give 75 % purples and 25 % whites." This does imply that there was a 3: 1 ratio between purple and white flowers, but it does not imply that de Vries did recognise the 3: 1 ratio. In the notes, de Vries did not make a mathematical representation as he did 4 years later. Therefore, it can be concluded that the 1896 notes do not show that de Vries conceptualised the Mendelian ratios or formulated the law of segregation in 1896. There was still a huge gap to fill between the implicit 3: 1 ratio in de Vries' 1896 notes and the Mendelian 3: 1 ratio derived from the law of segregation in 1900.

3.2.2 *Mendel and the Law of Segregation*

Many may still contend that had de Vries not recognised the Mendelian 3: 1 ratio in 1896, he would have recognised the ratio and formulated the law of segregation in the following years anyway, especially given his extensive experiments on plant hybrids in the 1890s.

I really doubt this. As I have argued, de Vries' Mendelian 3: 1 ratio was more than a statistical summary of the results of the hybridisation experiments. It also resulted from creative conceptualisation. More precisely speaking, the Mendelian 3: 1 ratio was impossible without the conceptions of dominance and of recessiveness. As I have shown in Sect. 2.4, the concept of dominance was first carefully articulated by Mendel (1866). And de Vries's use of dominance was clearly influenced by

Mendel's. In his pre-1900 publications, de Vries seldom used the terms "dominant" and "recessive". The first time that de Vries talked of dominance and recessiveness was in his 1900 papers.

In addition, it seems that de Vries did not recognise the universal regularity of the 3: 1 ratio until 1900. A clear case is his *Lychnis vespertina glabra* × *Lychnis diurna* experiment. De Vries began crossing *Lychnis vespertina glabra* with *Lychnis diurna* in 1892. In 1897 he reported that the distribution of the hairy and hairless flowers in F_2 generation in his *Lychnis vespertina glabra* × *Lychnis diurna* experiment was 2/3: 1/3.

> The distribution in 1894 was about 2/3 hairy and 1/3 hairless. (de Vries 1897, 89)

In 1899, de Vries recorded the distribution of hairy and hairless flowers as ¾: ¼.

> But whilst in 1893 all the hybrids had been hairy, this was no longer the case in 1894. Only about three-fourths were hairy, the rest hairless. I had 99 hairy and 54 hairless, in all 153 plants, and counted them in July at the commencement of flowering. (de Vries 1900b, 74)

In 1903 de Vries recorded this ratio as 73: 27.

> I record the 1892 crossing here, which gave rise to red-flowered hybrids in 1893. From their seeds, in the next summer, 116 plants bloomed red and 42 white, or 73% and 27%.[4] (de Vries 1903, 153)

The variation of the ratio is really interesting. In particular, it is a bit surprising to see that de Vries approximated 99: 54 to be 3: 1 ("3/4: 1/4") in 1899. Clearly, as Conway Zirkle (1968) and Malcolm Kottler (1979) indicate, 99: 54 is much closer to 2: 1 rather than 3: 1.

Two points can be made here. First, the variation of de Vries' record well reflects that in 1897 de Vries was not convinced that the ratio should be 3: 1. Thus, it further confirms my argument earlier that de Vries did not conceptualise the 3: 1 ratio in 1896.

Second, the discrepancy also suggests de Vries' changing view. In 1899, de Vries had not been fully convinced by the 3: 1 ratio. But in 1900 that de Vries began recognising the 3: 1 ratio and its general applicability.

To sum up, in 1900 de Vires made two remarkable changes. Firstly, he started using the phrases "dominant" and "recessive." Secondly, he began recognising the 3: 1 ratio in the F_2 generation.

But what made de Vries' sudden changes in 1900? I think that de Vries made a careful reading of Mendel's paper in late 1899 might be a best explanation. Otherwise it would be a miracle for de Vries to start talking of the 3: 1 ratio in terms of dominance and recessiveness suddenly in 1900. However, it should be noted that this does not imply that 1899 was the first time for de Vries to read Mendel.[5]

---

[4]"Ich führe hier die Kreuzung von 1892 an, welche im Jahre 1893 rothblühende Bastarde gab. Aus ihren Samen blühten im nächsten Sommer 116 Pflanzen roth und 42 weiss, oder 73 und 27%."

[5]There are different accounts of when and how de Vries first read Mendel's paper (de Vries 1900a, 85 n1; Roberts 1929, 323; Stomps 1954, 294).

Nevertheless, in late 1899 Mendel's paper made an impact on de Vries' work on hybridisation and heredity. Therefore, I argue that Mendel's influence was indispensable for de Vries' proposal of the law of segregation.

3.3 De Vries' Introduction of Segregation

Although Mendel's paper played a crucial role in de Vries' theorising of the law of segregation, it should not underrate de Vries' contribution. De Vries did not simply restate Mendel's writing. I shall argue that de Vries did make three steps forward beyond Mendel. Firstly, de Vries introduced Mendel's work into the study of heredity. Secondly, de Vries further developed Mendel's concepts of "dominant" and "recessive". Thirdly, de Vries introduced the idea of segregation.

3.3.1 From Mendel to Mendelism

As shown in Chap. 2, Mendel's work on hybrids was to study the development of hybrids and their progeny. However, de Vries had a different concern. His work on hybrids was to defend his theory of heredity experimentally: the theory of pangenesis (de Vries 1889). This purpose was clearly stated in the introductions of de Vries' papers (1900a, c).

> According to the principles which I have expressed elsewhere (Intracelluläre Pangenesis, 1889), the specific characters of organisms are composed of separate units. One is able to study, experimentally, these units either by the phenomena of variability and mutability or by the product of hybrids. (de Vries 1900c, 845, 1950, 30).
>
> According to pangenesis the total character of a plant is built up of distinct units... For many years this principle has represented the starting point for my investigations. Many important consequences can be deduced from it and may be tested experimentally. My experiments lie in part in the realm of variability and mutability and in part in that of hybridization. (de Vries 1900a, 83, 1966, 107).

Thus, de Vries' work (1900a, c, d) originated from his study of hybridisation in the 1890s: to test the theory of pangenesis experimentally. Moreover, in the conclusions, de Vries argued that his Mendelian law of segregation confirmed the theory of pangenesis.

> The totality of these experiments establishes the law of segregation of hybrids and confirms the principles that I have expressed concerning the specific characters considered as being distinct units. (de Vries 1900c, 847, 1950, 32)
>
> From these and numerous other experiments I draw the conclusion that the law of segregation as discovered by Mendel for peas finds very general application in the plant kingdom and that it has a basic significance for the study of the units of which the species character is composed. (de Vries 1900a, 84, 1966, 117).

It is worth noting that it is the first time in history that Mendel's work was used to study the problem of heredity. Of course, de Vries' Mendelism was substantially distinct from the early Mendelians'. For example, Bateson (1902) contended that Mendel's work provided a promising approach to the problem of heredity. A well-developed Mendelian theory of heredity would well account for both the "inward and essential nature" and the "outward and visible phenomena" of heredity. For de Vries, the significance of Mendel's work was still secondary. It was important for de Vries because it could be reformulated to provide a support of his theory of pangenesis. Nevertheless, de Vries was unequivocally the first to attempt to incorporate Mendel's work into the study of heredity.

3.3.2 From Activeness to Dominance

It is noticeable that de Vries interchangeably used the term "dominant" with "active (or visible)" on the one hand, and "recessive" with "latent" on the other hand in his 1900 papers.

> In the hybrid the simple differential character from one of the parents is accordingly visible or dominant while the antagonistic character is in the latent condition or recessive.[6] (de Vries 1900c, 845, 1950, 30).
>
> Of the two antagonistic characters, Mendel calls the one visible in the hybrid the dominating, the latent one recessive. (de Vries 1900a, 85, 1966, 111).

Alain Corcos and Floyd Monaghan (1985) contend that de Vries, following Mendel's usage, attributed "dominant" (or "visible") and "recessive" (or "latent") to the morphological trait, while Bert Theunissen (1994) argues that de Vries replaced "active/latent" with Mendel's "dominant/recessive" to designate the states of pangens. However, neither correctly reflects de Vries' subtle usage of "dominating/visible" and "recessive/latent" in 1900. Mendel mainly referred "dominant" and "recessive" to either the parental or hybrid trait with certain behaviour in the progeny. But for de Vries, the terms "dominant" and "recessive" were used to label two different pairs of things. On the one hand, de Vries referred these to the pair of morphological traits (*caractères, Merkmal*).

> In the hybrid the simple differential character [*caractère*] from one of the parents is accordingly visible or dominant while the antagonistic character [*caractère*] is in the latent condition or recessive. (de Vries 1900c, 845, 1950, 30)
>
> The antagonistic characters [*caractères*] ordinarily remain combined during all of the vegetative life, one dominant, the other latent. (de Vries 1900c, 845, 1950, 30)

---

[6]It is the paragraph in the paper that made Correns to realise that de Vries' already well knew Mendel's paper. In his paper (Correns 1900), Correns ironically indicated that de Vries used the same terms as Mendel in describing the paired traits was "a strange coincidence (*einen merkwürdigen Zufall*)."

> The dominating and the recessive traits [*Merkmal*] are shown to be constant in the progeny, … In this experiment they yielded an average of 77% with the dominating and 23% the recessive trait [*Merkmal*]. (de Vries 1900a, 88, 1966, 114)

On the other hand, de Vries referred those to the pair of hereditary characteristics (*Eigenschaften*), or qualities[7] (*qualités*).

> One can unify the whole of those results by supposing that the two antagonistic qualities [*qualities*], dominant and recessive, are distributed [mutually exclusively] in equal parts to the pollen just as to the ovules. (de Vries 1900c, 847, 1950, 32).
>
> Of the two antagonistic [characteristics[8] (*Eigenschaften*)], Mendel calls the one visible in the hybrid the dominating, the latent one recessive. (de Vries 1900a, 85, 1966, 111)
>
> The individuals d and d^2 have only the dominating [characteristics (*Eigenschaft*)], those of r and r^2 constitution possess only the recessive [characteristics (*Eigenschaft*)], while the dr plants are obviously hybrid. (de Vries 1900a, 86, 1966, 112)

Neither *Eigenschaften* (*or qualities*) nor *Merkmal* (or *caractères*) can be simply conflated. The word *Merkmal*, also used by Mendel, generally referred to what nowadays we call the morphological trait, while *Eigenschaften* was originally used by de Vries in his book *Intracellular Pangenesis* (de Vries 1889) to denote the hereditary property, which could be passed onto the next generation. Therefore, de Vries' usage of "dominating/recessive" was genuinely different from Mendel's.

What is more, it is worth noting that de Vries' interchangeable usage of "dominant/recessive" with "visible (active)/latent" was not trivial. In *Intracellular Pangenesis*, de Vries originally attributed the terms "active" and "latent" to two states of pangens. According to the theory of pangenesis, a pangen is the bearer of hereditary characteristics. Every hereditary characteristics, no matter in how many species it may be found, has its special kind of pangens. All living protoplasm is built up of pangens. In the nucleus every kind of pangen of the given individual is represented; the remaining protoplasm in every cell contains chiefly only those that are to become active in it. With the exception of those kinds of pangens that become directly active in the nucleus, as for example those that dominate nuclear division, all the others have to leave the nucleus in order to become active. But most of the pangens of every sort remain latent in the nuclei, where they multiply, partly for the purpose of nuclear division, partly in order to pass on to the protoplasm. This delivery always involves only the kinds of pangens that have to begin to function. During this passage they can be transported by the currents of the protoplasm and carried into the various organs of the protoplasts. In short, a pangen has two states: active and latent. When it is in the active state, it moves from the nucleus to the cytoplasm to manifest its

[7]De Vries used the word *Eigenschaften* in (de Vries 1900a) and *qualités* in (de Vries 1900d). It should be noted that in the book (de Vries 1889), de Vries used *Eigenschaften* to refer to hereditary characteristics or qualities.

[8]Evelyn Stern's translations of *Eigenschaften* (de Vries 1966) are inconsistent. It is translated as characteristics in some places, while it is translated as characters in others (see Appendix). In order to distinguish *Eigenschaften* from *Merkmal* (and *Charakter*), I find that characteristics is a better translation, which is also consistent with de Vries' (1889) usage of *Eigenschaften* to denote the hereditary quality in Gager's translation (de Vries 1910).

Table 3.1 Mendel's and de Vries' usages of "dominant" and "recessive"

	Mendel (1866)		de Vries (1889)	de Vries (1900a, c, d)
Morphological traits with a certain behaviour	Dominant/recessive (as a parental trait)	Dominant/recessive (as a hybrid trait)		
Morphological traits	Dominant/recessive (very occasionally)			Dominant (visible)/recessive (latent)
Hereditary characteristics				Dominant/ recessive
Pangens (unit)			Active/ latent	

characteristics. When in the latent state, it remains in the nucleus with its characteristics "latent". Therefore, de Vries' usage of "dominating/recessive" (1900a, c, d) was also different from his original usage of "active/latent" (1889).

Given Mendel's usage (1866) of "dominant/recessive" and de Vries' usage (1889) of "active/latent" (see Table 3.1), de Vries' interchangeable usage (1900a, c, d) can be seen as an incomplete attempt of incorporating Mendel's terminology. Since de Vries' purpose (1900a, c) was to test the principle that the character of an organism is built up of distinct units, it is easily to infer from de Vries' book (1889) that units should be construed as pangens, though he did not enunciate this point in the 1900 papers. Thus, it can be expected that if de Vries introduced the conceptions of "dominant" and "recessive" to refer to the morphological traits, there were corresponding hereditary characteristics. This is exactly what de Vries did in the 1900 papers. But there was a difficulty for de Vries to explain the pattern of inheritance of dominant/recessive traits in terms of units or pangens. De Vries was hesitant to conflate "dominant/recessive" with "active/latent" to denote the state of a pangen. All that de Vries (1900a, c, d) conclusively showed is that the pollen grains and ovules having one characteristics in the formation behaved in accord with the law of probability. This is why in the German paper de Vries was more modest[9] in the conclusion by arguing that the law of segregation "has a basic significance for the study of the units which the species character is composed" (de Vries 1966, 117). He well recognised that the principle that the specific characters of organisms are composed of units was not well established by his hybridising experiments, though at that time he must have been optimistic on that the law of segregation and his Mendelian analysis of the hybridising experiments would be very helpful to confirm the theory of pangenesis. Three years later, de Vries made a

[9]In the first French paper (de Vries 1900c), de Vries boldly concluded that his hybridising experiments "confirms that principles that I have expressed concerning the specific characters considered as being distinct units". (de Vries 1950, 32)

more comprehensive incorporation in the book (1903). Over one third of the book focused on "the Mendelian laws of segregation."

Eventually, as Meijer (1985, 223) indicates, "[de Vries] did not succeed in accommodating [Mendel's work] in to his already existing findings and views." Shortly after, de Vries abandoned his incorporating project by dismissing the significance of Mendel's work. As he wrote to Bateson on October 30th, 1901,

> ... [I]t becomes more and more clear to me that Mendelism [as the law of segregation] is an exception to the general rule of crossing. It is in no way the rule! It seems to hold only good in derivative cases, such as real variety-characters. (Provine 1971, 68).

Nevertheless, de Vries in his unsuccessful incorporation still developed the concept of dominance. It was an important step beyond Mendel, because de Vries was the first to use the term "dominant" to designate the hereditary property. Although he never explicitly referred dominant to the unit of the hereditary material, de Vries paved the road for the early Mendelians (e.g. Bateson 1902).[10]

3.3.3 From Correspondence to Segregation

It is evident that Mendel did not have the idea of segregation. In the law of composition of hybrid fertilising cells (LDT), Mendel only stated that the cell types corresponded to the morphological forms. He was implicit on the process of the combination of different cells in the fertilisation. Some may argue that Mendel's symbolic representation of LDT (i.e. $\frac{A'}{A'}+\frac{A'}{a'}+\frac{A'}{a'}+\frac{a'}{a'}=A+2Aa+a$) suggested a process of segregation. However, it should be noted that on the right-hand side of the equation, Mendel put it as $A + 2Aa + a$ rather than $AA + 2Aa + aa$. Mendel's formulation was very reasonable: Mendel referred A, a, and Aa to different morphological forms: dominant, hybrid, and recessive, while he referred $\frac{A'}{A'}$, $\frac{A'}{a'}$, $\frac{a'}{a'}$ to different cell-types. Mendel used $A + 2Aa + a$ to represent the ratio of the different morphological forms (1: 2: 1) in the F_2 generation. The reason that Mendel did not put it as $AA + 2Aa + aa$ was simply that there was no morphological form as AA or aa. It should also be noted that the fractions in the left-hand side of the equation did not refer to the numerical relation of cell-types or cells. Take the example of $\frac{A'}{A'}$. For Mendel, it just meant that in the fertilised cells both the egg and pollen cells were type A'. Similarly, $\frac{A'}{a'}$ meant that there were two cell-types fused in the fertilised cells: A' and a', and $\frac{a'}{a'}$ meant that both of the egg and pollen cells were type a'. It is incorrect to read $\frac{A'}{a'}$ as that there was a 1: 1 ratio of the number of the cells with cell-type A' and those with cell-type a'. Thus, the left-hand side of the equation (i.e. $\frac{A'}{A'}+\frac{A'}{a'}+\frac{A'}{a'}+\frac{a'}{a'}$)

[10]Thus, I oppose Darden's argument that "de Vries was guilty of not clearly distinguishing the units from the characters. He obviously meant that the units separate in the formation of pollens and ovules" (1976, 158). De Vries did have clear distinction between hereditary units and hereditary characteristics. Moreover, he even distinguished hereditary characteristics with morphological characters (see Table 3.1).

was about the composition of cell-types, while the right-hand side (i.e. $A + 2Aa + a$) about the ratio of the morphological forms. In other words, Mendel's symbolic formulation $\frac{A'}{A'}+\frac{A'}{a'}+\frac{A'}{a'}+\frac{a'}{a'}=A+2Aa+a$ did not suggest anything about segregation.

However, de Vries was explicit on the point that there was a process of segregation in the formation of pollen grains and ovules. He represented the process of segregation in the formation of pollen grains and ovules as $(d+r)(d+r)=d^2+2dr+r^2$ where d referred to the hereditary units with the dominant characteristics, and r referred to the hereditary units with the recessive characteristics (de Vries 1900c, 847). Thus, for de Vries, $(d + r)(d + r) = d^2 + 2dr + r^2$ was a mathematical representation of the process of segregation during generative period: When two hybrids were crossed, the hereditary units with the dominant characteristics and the same amount of the hereditary units with the recessive characteristics in the pollen cells segregated, and reunited with the hereditary units with the dominant characteristics and the hereditary units with the recessive characteristic in the ovule cells. The ratio of the hereditary units with the dominant characteristics only, the heredity units with both dominant characteristics and recessive characteristics, and the hereditary units with the recessive characteristics only was 1: 2: 1.

Thus, I argue that the difference between Mendel's and de Vries' symbolic representation clearly shows de Vries' novel contribution to the understanding of the mechanism of heredity at the beginning of the twentieth century. It was de Vries who first proposed the idea of segregation of the hereditary material in the fertilising period.

3.4 Conclusion

De Vries' contribution to genetics should not be overestimated or underestimated. In this chapter, I have argued that de Vries made three major contribution. First of all, de Vries introduced a Mendelian approach to the study of heredity. Secondly, de Vries developed the concept of dominance. Thirdly, de Vries proposed the idea of segregation. All of these were influential among the early Mendelians, especially Bateson. In addition, I have argued that Mendel's work played a vital role in de Vries' contribution. De Vries' reading of Mendel's paper helped him to conceptualise the 3: 1 ratio in terms of dominance and recessiveness. To sum up, both Mendel's work and de Vries' work were indispensable for the introduction of the Mendelian approach to the problem of heredity in the first decade of the twentieth century.

References

Bateson, William. 1902. *Mendel's Principles of Heredity: A Defence*. Cambridge: Cambridge University Press.

Bowler, Peter J. 1989. *The Mendelian Revolution: The Emergence of Hereditarian Concepts in Modern Science and Society*. London: The Athlone Press.

Corcos, Alain F., and Floyd V. Monaghan. 1985. Role of de Vries in the Recovery of Mendel's Work I. Was de Vries Really an Independent Discoverer of Mendel? *Journal of Heredity* 76 (3): 187–190.

Correns, Carl. 1900. G. Mendels Regel über das Verhalten der Nachkommenschaft der Rassenbastarde. *Berichte der Deutschen Botanischen Gesellschaft* 18 (4): 158–168.

———. 1966. G. Mendel's Law Concerning the Behavior of Progeny of Varietal Hybrids. In *The Origin of Genetics: A Mendel Source Book*, edited by Curt Stern and Eva R. Sherwood, translated by Leonie Kellen Piternick, 119–32. San Francisco, CA: W. H. Freeman and Company.

Darden, Lindley. 1976. Reasoning in Scientific Change: Charles Darwin, Hugo de Vries, and the Discover of Segregation. *Studies in History and Philosophy of Science* 7 (2): 127–169.

de Vries, Hugo. 1889. *Intracellulare Pangenesis*. Jean: Gustav Fischer.

———. 1897. Erfelijke Monstrositeiten in den Ruilhandel der Botanische Tuinen. *Botanisch Jaarboek* 4: 62–93.

———. 1900a. Das Spaltungsgesetz der Bastarde (Vorlaufige Mittheilung). *Berichte der Deutschen Botanischen Gesellschaft* 18 (3): 83–90.

———. 1900b. Hybridising of Monstroities. *Journal of the Royal Horticultural Society* 24: 69–75.

———. 1900c. Sur la Loi de Disjonction des Hybrides. *Comptes Rendus de l'Academie des Sciences (Paris)* 130: 845–847.

———. 1900d. Sur les Unités des Caractères Spécifiques et Leur Application à l'étude des Hybrides. *Revue Générate de Botanique* 12: 257–271.

———. 1903. *Die Mutationstheorie (II)*. Lepzig: Verlag von Veit & Comp.

———. 1910. *Intracellular Pangenesis*. Translated by C. Stuart Gager. Chicago, IL: The Open Court Publishing Co.

———. 1950. Concerning the Law of Segregation of Hybrids. Translated by Aloha Hannah. *Genetics* 35 (5, (2)): 30–32.

Vries, Hugo de. 1966. The Law of Segregation of Hybrids. In *The Origin of Genetics: A Mendel Source Book*, edited by Curt Stern and Eva R. Sherwood, translated by Evelyn Stern, 107–17. San Francisco, CA: W. H. Freeman and Company.

Gliboff, Sander. 2015. Breeding Better Peas, Pumpkins, and Peasants: The Practical Mendelism of Erich Tschermak. In *New Perspectives on the History of Life Sciences and Agriculture*, edited by Denise Philips and Sharon Kingsland, 395–413. Cham: Springer.

Kottler, Malcolm J. 1979. Hugo de Vries and the Rediscovery of Mendel's Laws. *Annals of Science* 36 (5): 517–538.

Meijer, Onno G. 1985. Hugo de Vries No Mendelian? *Annals of Science* 42 (3): 189–232.

Mendel, Gregor. 1866. Versuche über Pflanzenhybriden. *Verhandlungen des Naturforschenden Vereins Brünn* IV (1865) (Abhandlungen): 3–47.

———. 1966. Experiments on Plant Hybrids. In *The Origin of Genetics: A Mendel Source Book*, edited by Curt Stern and Eva R. Sherwood, translated by Eva R. Sherwood, 1–48. San Francisco, CA: W. H. Freeman and Company.

Monaghan, Floyd V., and Alain F. Corcos. 1986. Tschermak: A Non-Discoverer of Mendelism, I. An Historical Note. *Journal of Heredity* 77 (6): 468–469.

Olby, Robert Cecil. 1985. *Origins of Mendelism*. 2nd ed. Chicago, IL: The University of Chicago Press.

———. 1989. Rediscovery as an Historical Concept. In *New Trends in the History of Science*, edited by R. P. W. Visser, H. J. M. Bos, L. C. Palm, and H. A. M. Snelders, 197–208. Amsterdam and Atlanta: Rodopi.

Orel, Vitězslav. 1996. *Gregor Mendel: The First Geneticist*. Oxford: Oxford University Press.

Provine, William B. 1971. *The Origin of Theoretical Population Genetics*. Chicago, IL: The University of Chicago Press.

Quetelet, Adolphe. 1871. *Anthropométrie ou Mesure des différentes facultés de l'homme*. Brussell: C. Muquardt.

Rheinberger, Hans-Jörg. 1995. When did Correns Read Gregor Mendel's Paper? *Isis* 86 (4): 612–616.

Roberts, H.F. 1929. *Plant Hybridization before Mendel*. Princeton, NJ: Princeton University Press.

Simunek, Michal, Uwe Hoßfeld, and Olaf Breidbach. 2011. 'Rediscovery Revised' – The Coopertation of Erich and Armin von Tschermak-Seysenegg in the Context of the 'Rediscovery' of Mendel's Law in 1899–1901. *Plant Biology* 13 (6): 835–841.

———. 2012. 'Further Development' of Mendel's Legacy? Erich von Tschermak-Seysenegg in the Context of Mendelian-Biometry Controversy, 1901-1906. *Theory in Biosciences* 131 (4): 243–252.

Stamhuis, Ida H. 2015. Why the Rediscoverer Ended up on the Sidelines: Hugo de Vries's Theory of Inheritance and the Mendelian Laws. *Science & Education* 24 (1-2): 29–49.

Stamhuis, Ida H., Onno G. Meijer, and Erik J.A. Zevenhuizen. 1999. Hugo de Vries on Heredity, 1889–1903: Statistics, Mendelian Laws, Pangenes, Mutations. *Isis* 90 (2): 238–267.

Stomps, Theodoor Jan. 1954. On the Rediscovery of Mendel's Work by Hugo de Vries. *Journal of Heredity* 45 (6): 293–294.

Theunissen, Bert. 1994. Closing the Door on Hugo de Vries' Mendelism. *Annals of Science* 51 (3): 225–248.

von Tschermak, Erich. 1900a. Über Künstliche Kreuzung bei Pisum Sativum. *Berichte der Deutschen Botanischen Gesellschaft* 18 (6): 232–239.

———. 1900b. Über Künstliche Kreuzung bei Pisum Sativum. *Zeitschrift für das Landwirtschaftliche Versuchswesen in Oesterreich* 3: 465–555.

———. 1950. Concerning Artifical Crossing in Pisum Sativum. Translated by Aloha Hannah. *Genetics* 35 (5 (2)): 42–47.

Weiling, Franz. 1991. Historical Study: Johann Gregor Mendel 1822–1884. *American Journal of Medical Genetics* 40 (1): 1–25.

Zevenhuizen, Erik. 2000. Keeping and Scrapping: The Story of a Mendelian Lecture Plate of Hugo de Vries. *Annals of Science* 57 (4): 329–352.

Zirkle, Conway. 1968. The Role of Liberty Hyde Bailey and Hugo de Vries in the Rediscovery of Mendelism. *Journal of the History of Biology* 1 (2): 205–218.

Chapter 4
Weldon's Choice Reconsidered

> *Weldon's most sustained statement of [the] developmentalist, interactionist perspective on inheritance lies buried away in a manuscript entitled Theory of inheritance that he left unpublished and indeed uncompleted at his death in 1906.*
>
> (Jamieson and Radick 2017, 1263)

Abstract Along with Pearson, Weldon is best known as leading the Biometric school to resist the Mendelian approach, mainly developed by Bateson, to the study of heredity in the first decade of the twentieth century. Accordingly, the examinations of Weldon's work are typically framed within the context of the Mendelian-Biometrician controversy. This chapter revisits the significance of Weldon's work in the history of genetics by examining Weldon's view on inheritance and its development. Firstly, I critically review the traditional historiography of Weldon. Then, I sketch an outline of Weldon's later work on inheritance. Finally, I discuss the differences between Pearson's and Weldon's views on inheritance and science, and suggest a new reading of Weldon's work.

Keywords Weldon · Mendelism · Biometry · Pearson · Galton

4.1 The Mendelian-Biometrician Controversy

The introduction of Gregor Mendel's work into the study of heredity in 1900 was quickly followed up by an attempt to develop a Mendelian approach. With his associates, William Bateson tried to establish a new theory of heredity based on Mendel's work. The emergence of Mendelism led to its confrontation with Biometry, which was a popular approach to the study of heredity at the time. The Biometric approach was a statistical approach centred on Francis Galton's law of ancestral heredity, the basic idea of which was that the two parents contributed between them on average one-half of the total heritage of the offspring, the four grandparents one-quarter, and so on.

Y. Shan, *Doing Integrated History and Philosophy of Science: A Case Study of the Origin of Genetics*, Boston Studies in the Philosophy and History of Science 320,
https://doi.org/10.1007/978-3-030-50617-9_4

Along with Karl Pearson, W. F. R. Weldon is best known as leading the Biometric school to resist the Mendelian approach. However, the significance of Weldon in the history of genetics has not been fully appreciated for a long time until recently. There is an increasing interest in the reassessment of Weldon's legacy (e.g. Sloan 2000; Radick 2005, 2016a, b; Pence 2011; Jamieson and Radick 2013, 2017). Gregory Radick (2005, 2016a), for example, highlights that Weldon's unpublished work, *Theory of Inheritance*, contained an invaluable but overlooked project on a new theory of inheritance. Charles Pence (2011) points out that Pearson and Weldon had different motivations of their embrace of the statistical approach. Nevertheless, these recent examinations of Weldon's work are still to a great extent framed within the context of the Mendelian-Biometrician controversy. Pence, while highlighting the different motivations of Pearson and Weldon, still contends that Weldon and Pearson favoured the Biometry over Mendelism. Radick interprets Weldon's unfinished project as an attempt to develop a new Biometric theory against the Mendelian theory. This chapter revisits the significance of Weldon's work in the history of genetics by examining Weldon's view on inheritance and its development. In Sect. 4.2, I critically review the traditional historiography of Weldon. In Sect. 4.3, I sketch an outline of Weldon's later work on inheritance. In Sect. 4.4, I discuss the differences between Pearson's and Weldon's views on inheritance and science, and suggest a new reading of Weldon's work.

4.2 Weldon as a Biometrician

Traditionally, Weldon's work on inheritance is analysed and examined in the framework of the Mendelian-Biometrician controversy, in which Weldon is regarded as a leading figure of the Biometric School. This historiography consists of three central theses.

W1. With Pearson, Weldon led the criticisms of the Mendelian approach to the study of heredity.

W2. Like Pearson, Weldon favoured the statistical approach to the study of inheritance.

W3. Weldon, in *Theory of Inheritance*, aimed at a synthesis theory to render Mendelism as a special case in the Biometric framework.

Prima facie, all these theses seem to be well established. It is clear that Weldon was a critics of the Mendelian approach to the study of inheritance at the time. All Weldon's publications about inheritance in 1902–1904 were the criticisms of the Mendelian theory of inheritance (see Table 4.1).

Weldon's advocacy of the statistical approach is also clearly evidenced. In the editorial of the first issue of *Biometrika*, with two other co-founders of the journal, Pearson and Charles Davenport, Weldon provided an argument for the legitimacy of the statistical approach to the study of evolution and inheritance.

Table 4.1 Weldon's early publications on inheritance (1902–1904)

1902. "Mendel's Laws of Alternative Inheritance in Peas." *Biometrika* 1 (2): 228–54.
1902. "On the Ambiguity of Mendel's Categories." *Biometrika* 2 (1): 44–55.
1903. "Mr Bateson's Revisions of Mendel's Theory of Heredity." *Biometrika* 2 (3): 286–98.
1903. "Mendel's Principles of Heredity in Mice." *Nature* 68 (1750): 34.
1904. "Albinism in Sicily and Mendel's Laws." *Biometrika* 3 (1): 107–9.

(a) The statistical approach is fundamental to the study of the mass-phenomena of variation.
(b) The study of variation is essential to the study of heredity and selection.
(c) Therefore, the statistical approach is essential to the study of heredity and selection. (Weldon et. al. 1901a, 3)

Moreover, it seems that Weldon had supported the statistical approach consistently since 1890s.

> The question raised by the Darwinian hypothesis are purely statistical, and the statistical method is the only one at present obvious by which that hypothesis can be experimentally checked. (Weldon 1895, 381)
>
> [I]n certain aspects of biological research, biometry is an instrument which can aid us effectively in our gropings after truth. (Weldon et al. 1901b, 6)
>
> [I]f we attempt to describe our experience of any character of living things, . . . we can predict, with very considerable accuracy, the result of a long series. . .
>
> The method by which long series of variable results can be described, in such a way that mind can easily remember and form a useful picture of each series as a whole, is provided by the science of statistics. . ." (Weldon 1906, 96)

W3 is argued for by Robert Obly and Radick based on their reading of Weldon's unpublished manuscripts.

> Weldon's strategy by 1904 was to accommodate Mendelian heredity in a wider framework and to make it a special case. The effect of this approach was intended to preserve Galtonian heredity in the privileged position as the form of heredity upon which debate over the mechanism of evolution should be conducted. In that it could be deployed independently of any physiological hypothesis, Galtonian heredity was, in Weldon's opinion, superior to Mendelian heredity. . . [It is not surprising that] he produced a theory of heredity which effectively absorbed Mendelism into the larger embrace of Galtonian heredity. (Olby 1989, 316)
>
> Weldon's early death cut short the most promising interpretative effort on the behalf of ancestrian biometry. (Radick 2005, 37)

At first glance, these interpretations seem to be consistent with Weldon's own words.

> Mendel's work and Galton's are therefore in a sense complementary, the one dealing with the case in which selective mating is carried to its extreme limit among the ancestors of the stock observed, while the parents belong to distinct races, the other dealing with the stock produced by parents of a single race, in which selective mating is reduced to a minimum. . . (Weldon 1905, f.73.r)

However, I would argue that the traditional historiography of Weldon is problematic. In particular, putting Weldon into the context of the Mendelian-Biometrician confrontation is not very helpful to understand Weldon's view on inheritance in order to assess its significance. The Mendelian-Biometrician dichotomy fails to capture Weldon's nuanced understanding of inheritance. First of all, Weldon's anti-Mendelism does not imply his full embrace of Biometry. At the time, the Biometric theory was not the only alternative theory to the Mendelian one. As Yves Delage (1903) summarised, there were more than 30 theories of inheritance at the turn of the twentieth century. So, from a logical point of view, Weldon might have been supported any theory other than Mendelism given his opposition to it. Moreover, even in his first attack of Mendelism, Weldon clearly pointed out the limited scope of the Biometric theory of inheritance.

> The work of Galton (No. 13) and Pearson (No. 24) has given us an expression for the effects of blended inheritance which seems likely to prove generally applicable, although the constants of the equations which express the relation between divergence from the mean in one generation and that in another may require modification in special cases. Our knowledge of particulate or mosaic inheritance, and of alternative inheritance, is however still rudimentary, and there is so much contradiction between the results obtained by different observers, that the evidence available is difficult to appreciate. (Weldon 1902a, 228)

Thus, I argue that Weldon's anti-Mendelian stance does not imply Weldon's uncritical acceptance of Biometry. Secondly, W2 is not clear about to what extent Weldon favoured the statistical approach. Was Weldon's advocacy of the statistical approach exclusive? Did Weldon strongly believe that the statistical approach is the only approach to the study of heredity in *Theory of Inheritance*, as Pearson did? Thirdly, did Weldon in fact defend a Biometric theory of inheritance in *Theory of Inheritance*, as Olby and Radick argue? Before delving into these questions, I find it necessary to make a brief introduction to Weldon's later work on inheritance (1904–1906).

4.3 Weldon's *Theory of Inheritance*

Weldon's work on inheritance in the first decade of the twentieth century can be divided into two periods: 1901–1904 and 1904–1906. There are two significant differences between the works in these periods. First of all, all of the works in 1901–1904 were published, while the majority of the works in 1904–1906 were unpublished and unfinished (see Table 4.2). The only published works in 1904–1906 are "Inheritance in Animals and Plants" and "Current Theories of the Hereditary

Table 4.2 Weldon's Late Works on Inheritance (1904–1906)

MS. Theory of Inheritance (1904–1906)
Current Theories of the Hereditary Process (1905)
MS. "Individuality of the Chromosomes" (1905?)
MS. "The Hypothesis of the Individuality of the Chromosomes and Mendelism" (1905?)
"Inheritance in Animals and Plants" (1906)
"Note on the offspring of thoroughbred chestnut mares" (1906)
Correspondence. (1904–1905)

Process." "Inheritance in Animals and Plants" is a paper based on Weldon's lecture in Oxford, in August 1905, published in 1906 as a chapter in a book *Lectures on the Method of Science*, edited by T. B. Strong. "Current Theories of the Hereditary Process" consists of eight brief reports[1] of a series of Weldon's lectures at University College London (UCL) from 1904 to 1905,[2] published in *Lancet*. The rest of Weldon's works on inheritance in 1904–1906 are never published and currently stored at UCL Special Collections. Among Weldon's late unpublished manuscripts, *Theory of Inheritance* is probably his most systematic work on inheritance.

Secondly, Weldon's works in 1901–1904 were basically the criticisms of Mendelism, while those in 1904–1906, as I shall show, contained more constructive work,[3] such as an unfinished project of a new theory of inheritance. Thus, in order to have a fuller understanding of Weldon's view on inheritance, I shall at first make a brief summary of *Theory of Inheritance*.

4.3.1 The Outline

Not only is Weldon's *Theory of Inheritance* never published, but also it is unfinished in the sense that there are missing chapters, missing tables, missing diagrams, and missing references. There are ten chapters in the existing manuscript. There are two Chapter Is: Chapter I (A) and Chapter I (B). It is clear that Chapter I (B) is in fact an earlier draft of Chapter I (B). There are two Chapter IIs: One is entitled "The Theories of Inheritance based on Statistical Observation", while the other is noted as "Earlier Form of Chapter III." There is one Chapter III, entitled "Regeneration and Dominance." There are two Chapter IVs: Chapter IV (A) and Chapter IV (B). The latter is a continuation of the former, which is entitled as "Dominance and Mutilation in Embryo and Larvae." There are one Chapter V, one Chapter VII, and one Chapter VIII, in which Chapter VII is noted as "Earlier form: Alternative Chapter II". To sum up, there are in fact manuscripts of six chapters (see Table 4.3).

It is unclear how many chapters Weldon planned to write in total. However, the existing manuscript and the *Lancet* reports still provide us a rough idea of what might have been in Weldon's mind.[4] In the following sections, I shall examine the

[1]It should be noted that these reports were written by an unknown author.

[2]Weldon delivered a series of eight lectures on the subject of theories of heredity at UCL on 22/11/1904, 29/11/1904, 06/12/1904, 13/12/1904, 24/01/1905, 31/01/1905, 07/02/1905, and 14/02/1905 respectively.

[3]Note that Weldon's "Note on the offspring of thoroughbred chestnut mares" (1906b) was still basically a critique of Mendelism.

[4]As Olby (1989, 315) correctly indicates, the content of *Theory of Inheritance* and "Current Theories of Hereditary Processes" correspond to each other to a great extent. So it can be inferred that Weldon might have planned to write one or two chapters on chromosomes, given that there were two unpublished manuscripts on chromosomes (one is entitled "Individuality of the

Table 4.3 The Contents of *Theory of Inheritance* (Manuscript)

1	Chapter I. ("A") [files 1–39]	Chapter I. ("B") [files 40–63]
	Introduction (f.2.r – f.4.r)	(f.41.r – f.60.r)
	The Nature of Statistical Descriptions (f.5.r – f.14.r)	Comments: The earlier draft of Chapter I ("A")
	Correlation and Regeneration between two Varieties (f.14.r – f.22.r)	
	Appendix I. Methods of determining the Mean, and the moment coefficients about the mean (f.23.r – f.30.r)	
	Appendix II. Correlation and Regeneration (f.31.r – f.39.r)	
2	Chapter II. The Theories of Inheritance based on Statistical Observation [files 64–88]	Chapter VII. ("Earlier form: Alternative Chapter II") [files 220–250]
	Galton's Theory of Hereditary Transmission (f.65.r – f.72.r)	
	Mendel's Theory (f.73.r – f.70.r)	
	The Conceptions of Dominance and their Consequences (f.80.r – f.88.r)	
3	Chapter II. ("Earlier Form of Chapter III") [files 89–111]	Chapter III. Regeneration and Dominance [files 112–140]
4	Chapter IV. ("A") Dominance and Mutilation in Embryo and Larvae [files 142–167]	
	Chapter IV. ("B") ("from P II & be continuous at the end of Chapter IV A") [files 168–189]	
5	Chapter V. [files 190–219]	
6	Missing	
7	Missing	
8	Chapter VIII. [files 251–264]	
	Cuénot on Darbishire	

methodology, the aim, and the content of Weldon's work on inheritance in the manuscript.

4.3.2 *The Methodology of the Study of Inheritance*

In Chapter I (A), Weldon was very explicit on his methodology of the study of inheritance.

> The student of heredity has two main objects: the first is to discover what degree of stability is actually exhibited by the various races of animals or of plants, and to determine the extent

Chromosomes", and the other "The Hypothesis of the Individuality of the Chromosomes and Mendelism") and the fifth lecture of "Current Theories of Hereditary Processes" was about chromosomes.

> to which deviation from the average characters of parents or other ancestors is associated into deviation in their descendants; the second object is to acquire such knowledge of the changes which occur during the growth and maturation of the germ-cells, their fusion and subsequent development, as may serve to indicate the process by which the obscure relation between parents and filial characters is brought down. The first object is to make a purely descriptive statement of the actual relation between the visible bodily characters of living things and those of their ancestors or their descendants; the second is to learn the process to which this relation is due. These two objects are pursued by different methods, and as it happens they are generally pursued by different men, so that few attempts have been made to consider the learning of what are actually known concerning relation between the visible characters of parents and those of their offspring upon the possible interrelation of structural changes revealed by minute study of the germ-cells and of embryonic processes in germinal. (Weldon 1905, f.3.r)

For Weldon, the statistical approach was still important. The first step in the study of inheritance was to make an accurate statistical description of the pattern of inheritance. This well echoes Weldon's view on the statistical approach in his Oxford lecture on the method of biology in 1905.

> It is the first business of a scientific man to describe some portion of human experience as exactly as possible. It does not matter in the least what kind of experience he chooses to collect; his first business is to describe it. (Weldon 1906a, 81).

Nevertheless, it should be noted that the statistical approach, though important, was not the only approach to the study of inheritance. Weldon did not regard a purely statistical description of the pattern of inheritance as a complete theory of inheritance. In other words, what Weldon (1905, 1906a) argued for is in fact the priority of the statistical approach rather than the exclusivity of the legitimacy of the statistical approach. For Weldon, another important and indispensable task of the study of inheritance was to look for the mechanism of inheritance. As the title of the Lancet reports suggest, what Weldon was interested in is not only the pattern of inheritance, but also the process (or mechanism) of inheritance ("the hereditary process"). Thus, the statistical approach was clearly not enough for Weldon's study of the process of inheritance. In other words, Weldon was not a methodological monist. He also emphasised the significance of experimentation in the study of inheritance. For Weldon, experimentation was important in two ways. First, experimentation was complementary for the purpose of studying the mechanism of inheritance. In the manuscript of Chapters III, IV (A), IV (B), V, and VIII, Weldon (1905) extensively discussed various experiments on regeneration and embryonic development. Second, experimentation was complementary to the statistical approach for the purpose of making a description of the patterns of heredity.

> When [the evolutionist] cannot observe and measure in Nature, then he must experiment on "populations" within the laboratory. But few biological laboratories have the space or the resources needed for dealing with the vital changes of populations, still less do the means at the disposal of individuals suffice for carrying, out extensive experiments of this character. Much has been done and undoubtedly more will be done by the Marine Biological Laboratories for the study of mass-phenomena, but what is urgently needed is the establishment of a well-equipped Biometric Farm Laboratory, where breeding and survival experiments on large numbers could be carried out with ample room and care and, when necessary, for long periods. (Weldon et al. 1901b, 3)

It is worth noting that in 1901, inspired by Georg von Guita's work (1898, 1900), Weldon began his breeding experiments on Japanese waltzing mice with A. D. Darbishire to study the problem of heredity.[5] Although Darbishire conducted the experiments for the first 2 years, Weldon was responsible for the whole plan of the experiments and the statistical interpretation of the results and tool over the actual breeding arrangements after Darbishire's leave in 1903.

4.3.3 *The Aim of* Theory of Inheritance

Weldon also made the aim of the book very clear.

> It is the purpose of this essay first of all to describe the two principal theories of inheritance which have been faced upon dual comparison between the characters of living things and those of their ancestors or their descendants, and afterwards to see how the facts of development and regeneration as well as the facts of inheritance supports one theory or the other; but before we begin our proper work, it is necessary to say a little about the methods by which the companion between parental and filial characters can most constructively be made. (Weldon 1905, f.4.r)

The "two principal theories of inheritance" were introduced and discussed in Chapter II, namely "Galton's theory of hereditary transmission" and "Mendel's theory." It should be highlighted that Weldon's so-called "Galton's theory" was not the Biometric theory based on Galton's law of ancestral heredity (Galton 1889, 1897; Pearson 1898, 1903). Rather the theoretical core of this "Galton's theory" consisted of Galton's theory of hereditary determinant elements (1872a, b) and Galton's law of ancestral heredity. Two theories were about two objects of the study of inheritance respectively. Galton's theory of hereditary determinant elements[6] provided a theoretical framework to study the mechanism of inheritance, while Galton's law of ancestral heredity provided a mathematical framework to describe the pattern of inheritance.

In addition, Weldon's so-called "Mendel's theory" was not Mendel's theory of hybrid development (1866). It was rather the Mendelian theory of heredity, mainly developed by Bateson (1902). Thus, it seems to be more appropriate to name the two theories the Galtonian theory and the Mendelian theory. Accordingly, the aim of

---

[5]Froggatt and Nevin (1971, 21) suggests that the objective of Weldon's and Darbishire's breeding experiments on mice was more than a test of the Mendelian principles. Rather it was to formulate a general law of inheritance, which would embrace the entire spectrum of heredity such that "the Mendelians were merely working at one end of the scale, the biometricians somewhat further down" (Pearson 1908, 93). It is undeniable that the breeding experiments on mice might had been useful for Weldon's unfinished project on a general theory of inheritance. However, it seems too hasty to me to conclude that Weldon had such an aim in mind when he began experimenting mice in 1901, as the early publications on mice breeding (e.g. Darbishire 1902, 1903a, b, 1904; Weldon 1903) focused on the examination of the Mendelian principles.

[6]For a short summary of Galton's theory of hereditary determinant elements, see Sect. 4.3.4.

Weldon's *Theory of Inheritance* can be reinterpreted as an assessment of the Galtonian theory and the Mendelian theory. More precisely speaking, it purported to argue for the superiority of the Galtonian theory to the Mendelian theory. Throughout the manuscript, Weldon tried to show this point.

> Galton's theory of hereditary transmission has at least this advantage over Mendel's, that it takes all the known phenomena of inheritance into account, and endeavours to describe them all in terms of a single process. (Weldon 1905, f.84.r)
>
> While Galton has expressed those facts of blended inheritance for which Mendel's theory fails to account, there is nothing in his view of germinal constitution which is inconsistent with the facts of alternative inheritance observed by Mendel himself and by others after him. (Weldon 1905, f.85.r)
>
> These facts show us that we cannot regard the dominance as one of character over another in a Protozoa body as a permanent property due to something inherent in the nature of the determinants on which character depends; we cannot apply Mendel's view of dominance to any determinant elements we may conceive to exist in these animals: the facts of regeneration make it necessary to conceive the dominance of one determinant or of another as a temporary quality, depending not only on the nature of the determinant itself, but on the conditions under which it exists, so that in the same unit the dominant elements are sometimes of one kind and sometimes another; we are forced to take the same view of any determinants we may imagine to exist in a Protozoa as that which Galton holds concerning the nature of determinant elements in the higher animals. (Weldon 1905, f.122.r–f.123.r)
>
> Whatever may happen during the process of inheritance, it is clear that during the life and growth of simple individuals such as those we have examined, Mendel's conception of dominance as a property permanently belonging to the determinants of certain characters, whenever these are in the presence of certain others, is altogether inadequate; the tissues of these animals are neither "pure" in the sense that they contain only determinants representing a single character or group of characters, nor constant in the sense that some of the determinants they contain one of necessity dominant over the others; on the contrary, each tissue can be shown by experiment to be in the condition indicated by Galton's hypothesis, behaving as if it contained determinant elements which represent a large series of characters, any one or more of which can become either dominant or recessive, according to circumstances. (Weldon 1905, f.130.r–f.131.r)

4.3.4 The Theory

While Weldon aimed at arguing for the superiority of the Galtonian theory to the Mendelian theory, what Weldon's *Theory of Inheritance* provided us is more than a simple critical comparison. One of the distinctive features of the manuscripts is that Weldon attempted to take the studies of regeneration, of development and others into account. As Olby (1989, 315) indicates, "What to a modern reader is a remarkable feature of Weldon's treatment of heredity is the extent to which embryological data find a place in it." Thus, for Weldon, the study of inheritance was an interdisciplinary research across the various biological studies including embryology and cytology.

Another remarkable feature of Weldon's *Theory of Inheritance* is, as Radick (2016a, 156) correctly points out, that it focused on the problem of dominance.

Weldon's interest in the problem of dominance began in his first critical paper of Mendelism (1902a), in which he reformulated the Mendelian theory as two laws: the law of dominance and the law of segregation.[7] He was a critics of the Mendelian law of dominance and the Mendelian conception of dominance (see Weldon 1902a, b). His opposition to the Mendelian conception of dominance continued in his manuscript of *Theory of Inheritance*. That said, Weldon still found it necessary to have a better concept to describe and explain the phenomena of dominance. It should be highlighted that Weldon's conception of dominance was completely different from the Mendelian one, which referred to a property of hereditary elements. For Weldon, dominance referred to a process of how a hereditary determinant gave rise to a visible character.[8] This conception of dominance was developed from Galton's theory of hereditary determinant elements. The theory states that the appearance of any particular character is determined by hereditary elements under certain circumstances. Every such element exists either in an active, or a "dominant" condition, or in a "latent" condition. During development, these determinant elements become divided into an active or dominant group, which determines the characters of the body, and a latent or recessive group, which passes through the body developed from the germ without affecting its visible characters. When this body produces germ-cells, a sample of each group of determinants, both the dominant and the latent, is transmitted to every germ-cell. The latent elements, which pass from the germ out of which the parent was developed into the filial germs, without affecting the parental body, are derived from elements which have been dominant in some past ancestor, and may become dominant again in the filial at some later generation.

However, Galton's conception of dominance was quite rudimentary. As Weldon (1905, f.71.r) indicated, "[Galton] does not propose to offer any hypothesis concerning the nature of the determinant elements themselves, or of the condition on which their dominance or latency depends." Weldon suggested that there were different mechanisms of the manifestation of Galtonian dominance.

> In cases of blended inheritance, elements of different kinds, derived from each of a number of ancestors, appears to be usually dominant in the same individual; but in alternative inheritance we must assume that only elements of one kind can be dominant in the same body – the dominance of elements which determine one of the alternative characters being incompatible with the simultaneous dominance of those which represent the other in the germ. (Weldon 1905, f.72.r)

Moreover, Weldon anticipated a new conception of dominance, which might well account for the phenomena articulated by the Mendelian "dominance." As a result, the discussions on regeneration, reproduction, and development did not simply serve

[7]It should be highlighted that Mendel (1866) did not have the law of dominance or the law of segregation. This was Weldon's reformulation of Mendel's laws, which was criticised by the Mendelians. For example, Bateson (1909, 13) indicated that "[t]hose who first read of Mendel's work most unfortunately fell into the error of enunciating a 'Law of Dominance' as a proposition comparable with the discovery of segregation. Mendel himself enunciates no such law."

[8]Olby (1989, 315) provides a dispositional interpretation: "Weldon uses the word 'dominance' to mean that power which causes a determinant to be expressed."

as the evidential basis to make a comparative judgement of the Galtonian theory and the Mendelian theory, which was claimed by Weldon in the Introduction (1905, f.4.r). Rather, Weldon tried to accommodate the studies of regeneration, reproduction, development, and others into his Galtonian theory. In other words, Weldon's Galtonian theory was gradually developed into a new Weldonian theory of inheritance.

> The way in which the law of dominance is appreciated by selective mating has not yet been worked out, and a knowledge of the modifications which occur in particular cases may be expected to throw considerable light on the conditions by which the dominance or the latency of germinal elements is determined. (f.72.r Weldon 1905)
>
> Our knowledge of dominance has been much increased since Mendel wrote, partly by study of the effects produced by changing the normal relations of the tissues to each other, some of which we have described in Chapters II – IV, partly by Galton's work on inheritance, and partly by the more recent work inspired by Mendel's writings. The facts we now show us that the results which Mendel first showed to follow when certain races are crossed are not inconsistent with the belief that the elements transmitted to a hybrid are distributed among its gametes in the same way as those of a pure-bred individual, so that *we may conceive the relation between the constitution of the gametes and that of the individual which forms them to the same in all cases, and so express all the forms of inheritance in terms of one hypothesis*. (Weldon 1905, f.248.r)

Although Weldon never finished his project, three important conclusions can be drawn. First of all, the new Weldonian theory of inheritance was not a typical Biometric theory. It was clearly more than a purely statistical theory based on Galton's law of ancestral heredity. Second, it was not a simple synthesis of Galton's theory of hereditary determinant elements and the law of ancestral heredity. It encompassed the biological processes like regeneration and development, which were not covered by either of Galton's theories. Third, the Weldonian theory was not an assimilation of Mendelism into the Biometric framework. Neither the Mendelian approach nor the Biometric approach was adequate for Weldon. What Weldon finally would have produced is a new theory in which "both views, the Galtonian and the Mendelian, will be reconciled" ("Current Theories of the Hereditary Process" 1905, 732).

4.3.5 Summary and Remarks

To sum up, in *Theory of Inheritance*, Weldon tried to develop a new theory of inheritance, based on Galton's theory of hereditary determinants (1872a, b), the Biometric theory of inheritance (Galton 1889, 1897; Pearson 1898, 1903), and contemporary studies on regeneration, reproduction, development, embryology, and chromosomes (e.g. Hertwig and Hertwig 1887; Driesch 1897; Morgan 1902a, b), with a focus on the phenomenon of dominance. Thus, I argue that Weldon's advocacy of the statistical approach was not exclusive. Weldon contended that a statistical description of the pattern of inheritance was necessary, but he also needed other empirical methods to study the phenomena of inheritance. In other words, Weldon's advocacy of the statistical approach does not imply his full embrace of the Biometric theory of inheritance. Moreover, as I have argued, what Weldon worked on in *Theory of Inheritance* is in fact more ambitious than a better Biometric theory of inheritance. Therefore, the traditional historiography of Weldon is worth reconsidering.

4.4 Weldon, No Biometrician?

In this section, I discuss the differences of Weldon's and Pearson's views on the study of inheritance and science and suggest a new line of inquiry in the future study of Weldon.

4.4.1 Pearson vs. Weldon Reconsidered

Given Weldon's view on inheritance in *Theory of Inheritance*, I argue that Weldon's and Pearson's views differed in at least three significant aspects. Theoretically, Weldon and Pearson were substantially different. For Pearson, an ideal theory of inheritance would be a perfect statistical account of the pattern of inheritance, centred on Galton's law of ancestral heredity.

> If either that [Galton's] law, or its suggested modification be substantially correct, they embrace the whole theory of heredity. (Pearson 1898, 411)
>
> [I]f Darwinian evolution be natural selection with heredity, then [Galton's law of ancestral heredity] must prove almost as epoch-making to biologist as the law of gravitation to the astronomer. (Pearson 1898, 412)

However, what Weldon aimed at in his unfinished project is a theory which would be both descriptively adequate and explanatorily sophisticated. As Weldon (1905, f.3.r) highlighted, an ideal theory of inheritance consisted of "a purely descriptive statement of the actual relation between the visible bodily characters of living things and those of their ancestors on their descendants" and a mechanism of "the process to which this relation is due." Thus, for Weldon, a Biometric theory based on Galton's law was incomplete.

> The work of Galton (No. 13) and Pearson (No. 24) has given us an expression for the effects of blended inheritance which seems likely to prove generally applicable, although the constants of the equations which express the relation between divergence from the mean in one generation and that in another may require modification in special cases. Our knowledge of particulate or mosaic inheritance, and of alternative inheritance, is however still rudimentary, and there is so much contradiction between the results obtained by different observers, that the evidence available is difficult to appreciate. (Weldon 1902a, 228)
>
> [W]hen an explanation was sought of the mechanism or modus operandi of heredity, one passed... outside the domain of statistics and concrete facts and had to picture the invisible organic processes accompanying the growth and the reproduction of animals. ("Current Theories of The Hereditary Process." 1905, 42)

Such a theoretical disagreement was in fact rooted in a deeper methodological disagreement. For Pearson, an accurate description of the pattern of inheritance would be the full story that we need to know about inheritance. A mechanism of the process of inheritance was not necessary, though it could be helpful if it was served as a means to a better description.

> What I venture to think we require at present is not a hypothetical plasmic mechanics, but careful classifications of inheritance for the several grades of relationship, for a great variety of characters, and for many types of life. This will require not only the formation of records and extensive breeding experiments, but ultimately statistics and most laborious arithmetic... Such inventors are like planetary theorists rushing to prescribe a law of attraction for planets, the very orbital forms of which they have not first ascertained and described. Without the observations of TYCHO BRAHE, followed by the arithmetic of KEPLER, no NEWTON had been possible. The numerical laws for the intensity of inheritance must first be discovered from wide observation before plasmic mechanics can be anything but the purest hypothetical speculation. (Pearson and Lee 1900, 121)

In fact, Pearson's view on the methodology of the study of inheritance was a consequence of his view on the aim of science in general, which was neatly summarised in the Preface of his book *The Grammar of Science*.

> Step by step men of science are coming to recognise that mechanism is not at the bottom of phenomena, but is only the conceptual shorthand by aid of which they can briefly describe and resume phenomena. That all science is description and not explanation... (Pearson 1900, vii)

However, Weldon, though still arguing for the priority of the statistical approach, found it incomplete if taken as the only approach to science. As Weldon put it,

> It is the first business of a scientific man to describe some portion of human experience as exactly as possible. It does not matter in the least what kind of experience he chooses to collect; his first business is to describe it. (Weldon 1906a, 81).

It seems to many that this passage shows Weldon's support of the descriptive approach, but to me it clearly highlights the incompleteness of a purely descriptive approach. Weldon repeatedly emphasised that description was the "first business" of a scientist rather than the whole business of a scientist.

Furthermore, I argue that the methodological disagreement between Weldon and Pearson also reflects their different views on causation. It is not a completely new idea that Weldon and Pearson differed in their conceptions of causation. Pence (2011) suggests that Pearson had a positivist conception of causation in mind, while Weldon seemed to hold a probabilistic conception of causation. For Weldon, the main purpose of the statistical approach was "to permit us to retain (as much as is possible) the complexity of the biological world." (Pence 2011, 482).

I agree with Pence on the point that Pearson and Weldon differed substantially in the views on causation. Pearson's view on causation (1900) was fundamentally positivist, while Weldon's was clearly not. Nevertheless, I argue that Weldon's conception of causation, especially in his post-1904 works, is much closer to evidential pluralism than to the probabilistic theory of causation. Evidential pluralism, also called the Russo-Williamson thesis (2007), is an account of the epistemology of causation, maintaining that in order to establish a causal claim one has to have both correlative evidence and mechanistic evidence. As Jon Williamson recently elaborates,

> In order to establish a causal claim... one normally needs to establish two things: first, that the putative cause and effect are appropriately correlated; second, that there is some mechanism which explains instances of the putative effect in terms of the putative cause and which can account for this correlation. (Williamson 2019, 33)

Although evidential pluralism is primarily applied to biomedical sciences, it well illustrates Weldon's approach to the problem of dominance. In order to establish a causal claim about dominance, Weldon (1905, f.3.r) explicitly identified two tasks: the first was "to make purely descriptive statement of the actual relation between the visible bodily characters of living things and those of their ancestors on their descendants" (correlation), and the second was "to learn the process to which this relation is due" (mechanism). In other words, for Weldon, one had to have both the correlative evidence and the mechanistic evidence in order to determine what causes a dominant character. Therefore, it seems to me that Weldon, at least in 1904 or 1905, would had preferred evidential pluralism to a probabilistic theory of causation.

To sum up, the disagreement between Pearson and Weldon is shown to be more substantial than what is depicted in the traditional historiography, though they were close friends and intimate collaborators. In particular, when Weldon was writing *Theory of Inheritance*, he clearly diverged from Pearson theoretically, methodologically, and philosophically.

4.4.2 *Beyond Mendelism and Biometry*

The historiographical analysis of the development of genetics in the first decade of the twentieth century is overwhelmingly framed in the Mendelian-Biometrician controversy. Much has been discussed on the nature of the confrontation, the origin of the controversy, and the development of the debate (e.g. Froggatt and Nevin 1971; Farrall 1975; MacKenzie and Barnes 1975; Norton 1975; Barnes 1980; Roll-Hansen 1980; Olby 1989). However, such a framework is becoming less useful and fruitful. Focusing on the controversy too much leads to a failure of grasping a full picture of the development of genetics in that period, especially given that the controversy mainly took place in England between Englishmen. It has also been argued that the distinction between Mendelians and Biometricians is not even helpful to reflect the theoretical and methodological disagreements in the controversy. Some, say, Davenport, changed the stance in the controversy. Others, like Weldon, as well as G. Udny Yule (1902) and Darbishire (1906), attempted to find an alternative to the Biometric and the Mendelian theories. In other words, not all the contenders in the controversy can be simply classified into one of the two camps. The Mendelian-Biometrician dichotomy is particularly useless in Weldon's case. As I have argued, Weldon's work, especially those unpublished, cannot be simply read as a defence of Biometry. Unlike Pearson, Weldon was never fully committed to a purely statistical theory of inheritance. Nor did he regard Galton's law of ancestral heredity as the only exclusive hard core of an ideal theory of inheritance.

An important lesson from the significant differences between Weldon's and Pearson's views on inheritance and science is that it might be better to analyse and examine Weldon's work and its significance beyond the framework of the Mendelian-Biometrician controversy.

4.5 Conclusion

In this chapter, I have argued that a close reading of Weldon's *Theory of Inheritance* shows a different picture of Weldon in the history of genetics. He was clearly not a Biometrician in the same sense as Pearson was. He was not stubbornly committed to a Galtonian theory of heredity, centred on the law of ancestral heredity. His unfinished project was not a simple accommodation of Mendelism into Biometry. Rather Weldon was working on a general theory of inheritance, consisting of a mechanism of inheritance and a statistical account of the patterns of inheritance. It seems very clear to me that a better understanding of Weldon's work and its significance should be made by going beyond the framework of the Mendelian-Biometrician confrontation.

References

Barnes, Barry. 1980. On the Causal Explanation of Scientific Judgment. *Social Science Information* 19 (4/5): 685–695.

Bateson, William. 1902. *Mendel's Principles of Heredity: A Defence*. Cambridge: Cambridge University Press.

———. 1909. *Mendel's Principles of Heredity*. Cambridge: Cambridge University Press.

"Current Theories of the Hereditary Process". 1905. *The Lancet* 165 (4245): 42.

Darbishire, Arthur Dunkinfield. 1902. Note on the Results of Crossing Japanese Waltzing Mice with European Albino Races. *Biometrika* 2 (1): 101–104.

———. 1903a. Second Report on the Result of Crossing Japanese Waltzing Mice with European Albino Races. *Biometrika* 2 (2): 165–173.

———. 1903b. Third Report on Hybrids Between Waltzing Mice and Albino Races: On the Result of Crossing Japanese Waltzing Mice with 'Extracted' Recessive Albinos. *Biometrika* 2 (3): 282–285.

———. 1904. On the Result of Crossing Japanese Waltzing with Albino Mice. *Biometrika* 3 (1): 1–51.

———. 1906. On the Difference between Physiological and Statistical Laws of Heredity. *Memoirs and Proceedings of the Manchester Library and Philosophical Society* 50 (11): 1–44.

Delage, Yves. 1903. *L'hérédité et les Grands Problèmes de la Biologie Générale*. 2nd ed. Paris: Schleicher Frères.

Driesch, Hans. 1897. Studien über das Regulationsvermögen der Organismen 1. Von den Regulativen Wachsthums- und Differenzirungsfähigkeiten der Tubularia. *Archiv für Entwicklungsmechanik der Organismen* 5 (3): 389–418.

Farrall, Lyndsay A. 1975. Controversy and Conflict in Science : A Case Study – The English Biometric School and Mendel's Laws. *Social Studies of Science* 5 (3): 269–301.
Froggatt, Peter, and Norman C. Nevin. 1971. The 'Law of Ancestral Heredity' and the Mendelian-Ancestrian Controversy in England, 1889–1906. *Journal of Medical Genetics* 8 (1): 1–36.
Galton, Francis. 1872a. On Blood-Relationship. *Proceedings of the Royal Society* 20: 394–402.
———. 1872b. On Blood-Relationship. *Nature* 6: 173–176.
———. 1889. *Natural Inheritance*. London and New York: Macmillan & Company.
———. 1897. The Average Contribution of Each Several Ancestor to the Total Heritage of the Offspring. *Proceedings of the Royal Society of London* 61: 401–413.
Hertwig, O., and R. Hertwig. 1887. Über den Befruchtungs- und Teilungsvorgang des tierischen Eies unter dem Einfluß äußerer Agentien. *Jenaische Zeitschrift für Naturwissenschaft* 13: 120–241.
Jamieson, Annie, and Gregory Radick. 2013. Putting Mendel in His Place: How Curriculum Reform in Genetics and Counterfactual History of Science Can Work Together. In *The Philosophy of Biology: A Companion for Educators*, edited by Kostas Kampourakis, 577–95. Dordrecht: Springer.
———. 2017. Genetic Determinism in the Genetics Curriculum. *Science & Education* 26 (10): 1261–90.
MacKenzie, Donald A., and S. Barry Barnes. 1975. Biometriker versus Mendelianer. Eine Kontroverse und ihre Erklärung. In *Wissenschaftssoziologie: Studien und Materialien*, edited by Nico Stehr and René König, 165–196. Wiesbaden: VS Verlag für Sozialwissenschaften.
Mendel, Gregor. 1866. Versuche über Pflanzenhybriden. *Verhandlungen des Naturforschenden Vereins Brünn* IV (1865) (Abhandlungen): 3–47.
Morgan, Thomas Hunt. 1902a. Experimental Studies of the Internal Factors of Regeneration in the Earthworm. *Archiv für Entwicklungsmechanik der Organismen* 14 (3–4): 562–591.
———. 1902b. The Reflexes Connected with Autotomy in the Hermit-Crab. *American Journal of Physiology* 6 (5): 278–282.
Norton, Bernard J. 1975. Biology and Philosophy: The Methodological Foundations of Biometry. *Journal of the History of Biology* 8 (1): 85–93.
Olby, Robert Cecil. 1989. The Dimensions of Scientific Discovery: The Biometric-Mendelian Debate. *The British Journal for the History of Science* 22 (3): 299–320.
Pearson, Karl. 1898. Mathematical Contributions to the Theory of Evolution. On the Law of Ancestral Heredity. *Proceedings of the Royal Society of London* 62: 386–412.
———. 1900. *The Grammar of Science*. 2nd ed. London: Adam and Charles Black.
———. 1903. The Law of Ancestral Heredity. *Biometrika* 2 (2): 211–229.
———. 1908. On a Mathematical Theory of Determinantal Inheritance, from Suggestions and Notes of the Late W. F. R. Weldon. *Biometrika* 6 (1): 80–93.
Pearson, Karl, and Alice Lee. 1900. "Mathematical Contributions to the Theory of Evolution. VIII. On the Inheritance of Characters Not Capable of Exact Quantitative Measurement. Part I. Introductory. Part II. On the Inheritance of Coat-Colour in Horses. Part III. On the Inheritance of Eye-Co." *Philosophical Transactions of the Royal Society of London.* Series A, Containing Papers of a Mathematical or Physical Character, 195: 79–150.
Pence, Charles H. 2011. 'Describing Our Whole Experience': The Statistical Philosophies of W. F. R. Weldon and Karl Pearson. *Studies in History and Philosophy of Biological and Biomedical Sciences* 42 (4): 475–485.
Radick, Gregory. 2005. Other Histories, Other Biologies. *Royal Institute of Philosophy Supplement* 80 (56): 3–47.
———. 2016a. Presidential Address: Experimenting with the Scientific Past. *The British Journal for the History of Science* 49 (2): 153–172.
———. 2016b. Teach Students the Biology of Their Time. *Nature* 533 (7603): 293–293.
Roll-Hansen, Nils. 1980. The Controversy between Biometricians and Mendelians: A Test Case for the Sociology for Scientific Knowledge. *Social Science Information* 19 (3): 501–517.

Russo, Federica, and Jon Williamson. 2007. Interpreting Causality in the Health Sciences. *International Studies in the Philosophy of Science* 21 (2): 157–170.
Sloan, Philip R. 2000. Mach's Phenomenalism and the British Reception of Mendelism. *Comptes Rendus de l'Académie des Sciences – Series III – Sciences de la Vie* 323: 1069–1079.
von Guita, Georg. 1898. Versuche mit Kreuzungen von Verschiedenen Rassen der Hausmaus. *Bericht der Naturforschenden Gesellschaft zu Freiburg* 10: 317–332.
———. 1900. Versuche mit Kreuzungen von Verschiedenen Rassen der Hausmaus. *Bericht der Naturforschenden Gesellschaft zu Freiburg* 11: 131–138.
Weldon, Walter Frank Rapheal. 1895. Remarks on Variation in Animals and Plants. To Accompany the First Report of the Committee for Conducting Statistical Inquiries into the Measurable Characteristics of Plants and Animals. *Proceedings of the Royal Society of London* 57: 379–382.
———. 1902a. Mendel's Laws of Alternative Inheritance in Peas. *Biometrika* 1 (2): 228–254.
———. 1902b. On the Ambiguity of Mendel's Categories. *Biometrika* 2 (1): 44–55.
———. 1903. Mendel's Principles of Heredity in Mice. *Nature* 68 (1750): 34.
———. 1905. *Theory of Inheritance (Unpublished)*. London: UCL Library.
———. 1906a. Inheritance in Animals and Plants. In *Letures on the Method of Science*, edited by Thomas Banks Strong, 81–109. Oxford: Clarendon Press.
———. 1906b. Note on the Offspring of Thoroughbred Chestnut Mares. *Proceedings of the Royal Society B* 77 (519): 394–398.
Weldon, Walter Frank Rapheal, Karl Pearson, and Charles B. Davenport. 1901a. Editorial: The Scope of Biometrika. *Biometrika* 1 (1): 1–2.
———. 1901b. Editorial: The Spirit of Biometrika. *Biometrika* 1 (1): 3–6.
Williamson, Jon. 2019. Estabishing Causal Claims in Medicine. *International Studies in the Philosophy of Science* 32 (1): 33–61.
Yule, G. Udny. 1902. Mendel's Laws and Their Probable Relations to Intra-Racial Heredity. *New Phytologist* 1 (10): 222–238.

Part II
Integrated HPS

In Part I, I have argued that Mendel's conception of development was influenced by Gärtner, and his paper (1865) could not be interpreted as a study of heredity. I argued that in his introduction of Mendel's work to the study of heredity, de Vries developed Mendel's conception of dominance and proposed the law of segregation. I also argued that in his unpublished manuscript, Weldon was working on a new theory of inheritance, surpassing both the Mendelian and the Biometric theories. All these arguments contribute to a revised historiography of the origin of genetics: Mendel's work on hybrid development was introduced to the study of heredity in 1900 by de Vries, Correns, and Tschermak. Soon after, Bateson and his associates enthusiastically developed a Mendelian theory of heredity, which was resisted and criticised by many of their contemporaries. Among the critics, Weldon began working on a new theory of inheritance. Thus, an important task for integrated HPS is to make sense of this historiography. In this part, I introduce a new method of integrated HPS, namely, the exemplar-based approach. In Chap. 5, I propose an exemplar-based analysis of the origin of genetics. In Chap. 6, motivated by the exemplar-based approach, I develop and defend a new functional account of the progress in the history of genetics. Chapter 7, I argue for an exemplar-based explanation of the long neglect of the significance of Mendel's work for the study of heredity.

Chapter 5
Exemplarising the Origin of Genetics

> Mendel's paper illustrates an exemplar for how to set up an empirical practice.
>
> (Moss 2003, 23)

Abstract Understanding the origin of genetics has been a persistent problem in the history and philosophy of biology: Is Mendel the founder of genetics? If so, in what sense? What was Mendel's contribution to the origin of genetics? What role did Mendel's work play in the "rediscovery" in 1900? What was the contribution made by the "rediscoverers" to the origin of genetics? What is a best way to analyse and interpret the origin of genetics, from a philosophical point of view? This chapter provides a new philosophical account of the early development of genetics. I begin with a critical review of the theory-based analyses of the origin of genetics. Then inspired by the Kuhnian analyses, I develop an exemplar-based account of the origin of genetics based on a new interpretation of exemplar.

Keywords Exemplar · Exemplary practice · Mendel · Origin of genetics

5.1 Introduction

As I have discussed in the previous chapters, the historiography of genetics today has been radically revised. It is now a consensus that Gregor Mendel's concern was the development of pea hybrids rather than the problem of heredity in general. It is also accepted that the great "rediscovery of Mendel's work" in 1900 was in fact the introduction of Mendel's work into the study of heredity, though Hugo de Vries, Carl Correns, and Erich von Tschermak, the "rediscoverers" of Mendel's work in 1900, all differed in their research problems. Now I am turning to an integrated HPS question concerning the origin of genetics: What is a best way to analyse and interpret the origin of genetics, from a philosophical point of view? This chapter

Y. Shan, *Doing Integrated History and Philosophy of Science: A Case Study of the Origin of Genetics*, Boston Studies in the Philosophy and History of Science 320,
https://doi.org/10.1007/978-3-030-50617-9_5

provides a new account of the early development of genetics. In Sect. 5.2, I criticise the theory-based analyses of the origin of genetics. In Sect. 5.3, I review the Kuhnian analyses. In Sect. 5.4, I introduce a new method of integrated HPS, namely, the exemplar-based approach. In Sect. 5.5, I propose an exemplar-based account of the origin of genetics.

5.2 The Theory-Based Accounts of the Origin of Genetics

Traditionally, philosophers used to understand and analyse the origin of genetics in terms of theories. Michael Ruse's succinct summary is quite representative.

> Mendel's own work, as is well known, went practically unnoticed for thirty years. However, after its rediscovery at the beginning of [the 20th] century, a theory of heredity based on his ideas was developed in great depth and at a rapid speed. (Ruse 1973, 12)

The origin of genetics from Mendel to Mendelian genetics, for many philosophers like Ruse, is basically a process of the development of a theory. Accordingly, a philosophical analysis of the origin of genetics typically starts with identifying a central explanatory theory. Then it analyses its theoretical components (say, concepts and principles) in different periods, details how they can be applied to explain the phenomena, reconstructs how they develop and are justified, and explores the strategies for theoretical changes. Lindley Darden (1991), for example, makes such a detailed analysis of the development of the theoretical components in the history of genetics. She regards the works of early Mendelians, including de Vries (1900a, 1900b, 1900c), Correns (1900), William Bateson (1902, 1909), and the works of the Morgan School (e.g. Morgan et al. 1915; Morgan 1926) as the different versions of the theory of Mendelian genetics. She identifies theoretical components of theory of Mendelian genetics and examines their changes in the period 1900–1926. Thus, for Darden, the origin of genetics is a process of the development of the theory of Mendelian genetics (Table 5.1).

Philip Kitcher's account of the origin of genetics (1984, 1989) is also clearly theory-based. For him, the theoretical variations in the history of genetics can be formulated as the different explanatory schemata aiming at the problem of the transmission of traits. For example, the theory of Mendelian genetics in 1900 can be formulated in the following schematic argument:

(1) There are two alleles A, a. A is dominant, a recessive.
(2) AA and Aa individuals have trait P, aa individuals have trait P′.
(3) The genotypes of the individuals in the pedigree are as follows: i_1 is G_1, i_2 is G_2, . . .,i_N is G_N.
(4) For any individual x and any alleles y, z if x has yz then the probability that x will transmit y to any one of its offspring is 1/2.
(5) The expected distribution of progeny genotypes in a cross between i_j and i_k is D; the expected distribution of progeny genotypes in a cross . . .

Table 5.1 Darden's summary (1991) of the theoretical changes in Mendelian genetics between 1900 and 1926

Theoretical Knowledge	1900–1903	1926
The unit of inheritance	Units-characters	Genes
	An organism is to be regarded as composed of separable units/characters.	Genes cause characters.
		One gene may cause one one character.
		Multiple factors may interact in the production of one character.
		One gene may affect many characters.
Multiple alleles	In varieties of organisms, the traits by which they differ are antagonistic or differentiating pairs of characters.	In any one organism, genes occur in pairs.
		In a population, multiple allelomorphs for a character occasionally occur.
Connection with cytology	The connections between generations are the germ cells.	Genes are transmitted from parents to offspring in the germ cells.
Dominance-Recessiveness	In a hybrid formed by crossing parents that differ in a single pair of characters, there is some difference such that one character dominates over the over; thus, the character in the hybrid resembles one but not the other of the parents.	The distinction between dominance and recessiveness was deleted.
Mendel's law of segregation	In the formation of germ cells in a hybrid produced by crossing parents that differed in a single pair of characters, the parental characters segregate or separate, so that the germ cells are of one or other of the pure parental types.	Parental genes are not modified as a result of being together in a hybrid; no new kinds of hybrid genes form.
		In the formation of germ cells of a hybrid, paired parental genes segregate so that the germ cells have one or the other of a given pair.
	The two different types of germ cells form in approximately equal numbers.	The two different types of germ cells are formed in equal numbers.
	When two hybrids are fertilised, the differing types of germ cells combine randomly.	When two similar hybrids are cross-fertilised, the differing types of germ cells combine randomly.
Mendel's law of independent assortment	In the formation of germ cells in a hybrid produced by crossing parents that differed in two or more traits, the parental factors segregate or separate so that the germ cells are of all possible pure parental combinations.	Genes are found in linkage groups; groups occur in corresponding pairs.
	The different types of germ cells are formed in equal numbers.	Genes of different linkage group assort independently.
	When two hybrids are fertilised, the differing types of germ cells combine randomly.	Usually genes in the same linkage group are inherited together; at times, however, an orderly

(continued)

Table 5.1 (continued)

Theoretical Knowledge	1900–1903	1926
		interchange, called "crossing-over", takes place between allelomorphs in corresponding linkage groups.
		Genes in a linkage group are arranged linearly with respect to each other.
		The frequency of crossing-over can be used to calculate the order and relative positions of the respective genes in linkage groups if such disturbing factors as double cross-overs and interference of one cross-over with another are taken into account.
Mutation	No concept "mutation".	Genes occasionally mutate and then cause a different character.
		Mutation does not alter their linear relation to other genes in the linkage group.

(6) The expected distribution of progeny phenotypes in a cross between i_j and i_k is E; the expected distribution of progeny phenotypes in a cross ...[1] (Kitcher 1989, 439)

Similarly, the theory of Mendelian genetics in the period 1902–1910 and T. H. Morgan's theory of the gene in 1926 can be reformulated as two different schematic arguments (Kitcher 1989, 440–41). In short, Kitcher characterises the origin of genetics as the development of a theory in terms of explanatory schemata (or patterns of reasoning). These theory-based accounts of the origin of genetics are deeply rooted in a theory-driven approach in the philosophy of science, which is perfectly summarised by C. Kenneth Waters:

> Philosophers (perhaps I should say we) typically analyze scientific knowledge by identifying central explanatory theories. Then for each theory, we analyze its central concepts and principles (or laws), detail how it can be applied to explain the phenomena, reconstruct how it is justified, explore how it might be further developed or how its explanatory range might be extended (the so-called 'research program'), and consider how it should be interpreted (for example, instrumentally or realistically). (Waters 2004, 784)

However, such a theory-driven approach has been heavily criticised by many philosophers since the 1980s. The biggest problem is that the theory-driven approach overlooks the significance of experimentation and other non-theoretical activities in

[1]Note that (1), (2), (3), and (4) are premises; (5) is obtained from (3) and (4) by using the principles of probability; (6) is derived from (2) and (5).

Table 5.2 The main theoretical changes from Mendel to Bateson

The version of the theory of heredity	The cellular factor relating to trait	Relation of cellular correspondent-trait	Applicability of Mendel's Law
Mendel's version	Types of cell	Correspondence	Pisum and Phaseolus
De Vries' version	Characteristics	Determination	All true hybrids
Correns' version	Anlagen	Determination	Some hybrids
Bateson's version	Allelomorphs	Determination	Certain phenomena of alternative inheritance

science. (*The Problem of Practice*) As Ian Hacking (1983, 149) points out, "Philosophers of science constantly discuss theories and representation of reality, but say almost nothing about experiment, technology, or the use of knowledge to alter the world." Many (e.g. Hacking 1983; Cartwright et al. 1995) feel dissatisfied with this approach. One lesson that can be learnt from these criticisms is that the history of science, even if from a philosophical/intellectual perspective, should be much more than a history of scientific theory.

Accordingly, the theory-based analysis of the origin of genetics is also problematic. Consider the period from 1865–1902. From the theory-based viewpoint, given that Mendel's work (1866) was about development of hybrids in their progeny, it might be natural to argue that Mendel's major contribution is the proposal of a theory of hybrid development. In 1900 de Vries developed Mendel's theory by extending its applicability and refining the law. In the same year, Correns also reformulated Mendel's theory in terms of anlagen (Table 5.2).

It can be argued that there were two significant modifications to Mendel's theory made by de Vries and Correns. One is that the correspondence between kinds of cell and morphological traits in Mendel's work was replaced with a kind of determination in de Vries' and Correns' work. The other is that the interfield connection of the theory was advanced. Mendel's "types of cell" in fertilisation was reconceptualised as "characteristics" in the formation of pollen and ovules (de Vries 1900a, 1900b, 1900c) or "anlagen" in the process of the fusion of the reproductive nuclei (Correns 1900). Inspired by de Vries' and Correns' work, Bateson (1902) further developed the theory by refining the applicability and key concepts. Thus, the "essence" of the origin of genetics from Mendel to Bateson is characterised as a process of the development of a theory.

However, this theory-based characterisation is highly problematic. Firstly, if the "essence" of the origin of genetics is depicted as the development of a theory, it is extremely difficulty to identify such a theory. As I have shown in Chaps. 2 and 3, Mendel's work was not about heredity, while de Vries' and Correns' concerns were not about the development of hybrids in their progeny. De Vries, Correns, and

Tschermak in their 1900 papers had all different concerns. In particular, it is difficult to maintain that they proposed their theories explicitly, especially in Tschermak's case. It has been shown that Tschermak (1900a, 1900b) attempted to incorporate Mendel's analysis of the progeny of hybrids with his theory of regular differential valency of traits, though he did not have any sophisticated formulation in 1900 (e.g. Simunek et al. 2011, 2012). Thus, it is not obvious that there is a linear development of a theory from Mendel (1866) to Tschermak (1900a, 1900b). Correns (1900) aimed to test Mendel's work on *Pisum* experimentally, and ended up by proposing a Mendelian Rule. De Vries (1900a, 1900b, 1900c) proposed the law of segregation and intended to use it to support his theory of pangenesis. So, it is not historically accurate to summarise de Vries' and Correns' work as the development or revision of Mendel's theory of hybrid development, given that their concerns were not the development of hybrids in their progeny. In other words, there is a problem of identification. If we understand the origin of genetics as the development of a theory, there is a difficulty of identifying *the theory* of genetics, or the essence of the theory of genetics in that period.

Secondly, even if Correns' Mendelian Rule and de Vries' law of segregation can be construed as revised versions of Mendel's theory, such a theory-driven analysis of the origin of genetics fails to reflect the radical change of the subject of the theories. Accordingly, some more complicated problems occur: What makes Mendel's theory of hybrid development and de Vries' theory different versions of a theory? What is the connection between Mendel's and de Vries' theories? An adequate theory-based analysis of the origin of genetics has to articulate the change from Mendel's theory of hybrid development to de Vries' theory of heredity. There is much more to be done if one tries to defend the view that there was a development of the theory of Mendelian genetics from Mendel to Bateson.

Thirdly, the theory-driven analysis of the origin of genetics from Mendel to the "rediscoverers" faces the problem of practice. As I shall discuss in detail in Sect. 5.3, Mendel's work was much more than a theoretical construction, and the work of "rediscoverers" was heavily influenced by the non-theoretical aspect of Mendel's. The non-theoretical aspect of the origin of genetics is missing and largely neglected from the theory-driven analysis.

Of course, these objections are not decisive. Darden might accept that there is a problem of identification in the period 1865–1902, and would argue that this is the reason why her account of the origin of genetics starts from 1900 rather than 1865. Kitcher would have argued that his account well resolves the three problems. Firstly, he identifies the essence of the theory of genetics with the problem of transmission of traits (the pedigree problems). All of Mendel's, the Mendelian, and Morgan's theories are different patterns of reasoning in order to solve the problem of transmission of traits. Secondly, the theoretical variations can be characterised as the revisions of the premises of the schematic arguments. Thirdly, his analysis highlights the significance of the non-theoretical elements in the history of genetics. His concept "the practice of classical genetics" encompasses accepted statements as well as questions, patterns of reasoning, and experimental procedures and methodologies.

Table 5.3 Mendel's "Law(s)" between 1900 and 1926

<table>
<tr><th>Source</th><th colspan="2">Name</th></tr>
<tr><td>De Vries (1900a, 1900b, 1900c)</td><td colspan="2">The law of segregation</td></tr>
<tr><td>Correns (1900)</td><td colspan="2">Mendel's rule</td></tr>
<tr><td>Tschermak (1900a)</td><td colspan="2">The principle</td></tr>
<tr><td>Davenport (1901)</td><td colspan="2">Mendel's law of alternative inheritance</td></tr>
<tr><td rowspan="2">Weldon (1902)</td><td rowspan="2">Mendel's law of alternative inheritance</td><td>The law of dominance</td></tr>
<tr><td>The law of segregation</td></tr>
<tr><td>Bateson (1902, 1909)</td><td colspan="2">Mendel's law</td></tr>
<tr><td rowspan="2">Morgan (Morgan et al. 1915; Morgan 1926)</td><td colspan="2">Mendel's first law (or the law of segregation)</td></tr>
<tr><td colspan="2">Mendel's second law (or the law of independent assortment)</td></tr>
</table>

Unfortunately, these responses are not fully satisfactory. Darden's Mendel-free account of the origin of genetics leads to two further questions. Firstly, it fails to provide an adequate assessment of Mendel's contribution. Even if Mendel's work is excluded from a theory-based account of the origin of genetics, there is still an inevitable task to account for Mendel's contribution, given his great impact on de Vries', Correns', Tschermak's, and Bateson's work. Secondly, even if the analysis focuses on the period 1900–1926, there is still a problem of identification. It is clear there is no such a theoretical consensus among the so-called Mendelians in the period 1900–1903 summarised by Darden. Thus, as Staffan Müller-Wille and Vitězslav Orel (2007, 212) correctly point out, "Mendelians did not share one particular, and certainly not a particular [...] theory of inheritance." For example, as Table 5.3 shows, there was no consensus on the formulation of Mendel's law in the 1900s. Nor was the concept "unit/character" commonly used among early Mendelians. Thus, Darden's treatment fails to reflect the theoretical differences among Mendelians in the period 1900–1903.

Kitcher's potential response is historically flawed. Firstly, as I have argued in Chap. 2, Mendel's work was not to solve the problem of transmission. It is incorrect to regard Mendel's theory of hybrid development as an explanatory schema to solve the pedigree problem. Secondly, Kitcher's analysis pushes philosophers to hold an oversimplified understanding of the development of Mendelian genetics. A consequence is that the practice of Mendelian genetics aims to look for a theory with a great explanatory power. It is misleading and ahistorical. It is a fact that neither Mendel's study on *Pisum*[2] or Morgan's study on *Drosophila* initially aimed to look for a theory to explain the phenomena of inheritance. It is also evident, as Waters (2004) shows, that there are many significant contributions (e.g. the investigation of the crossover modifier C_{IIIB}) by classical geneticists, which did not aim at applying, extending or improving the theory of Mendelian genetics.

[2]For more discussion, see Chap. 2.

Now I can conclude that the theory-driven analyses of the history of genetics are problematic in the sense that both the theory-driven approach and its underlying assumption are problematic. Thus, it seems to be a good time to develop an alternative approach to analysing and interpreting the origin of genetics.

5.3 The Kuhnian Accounts

In contrast to philosophers, historians seem to prefer a Kuhnian account of the origin of genetics. On the basis of his reinterpretation of Mendel's work, Augustine Brannigan (1979) adopts Thomas Kuhn's conceptual framework to reassess Mendel's contribution:

> [I]n 1866 Mendel's work figured as normal science in the hybridist tradition, while in 1900 the revival of Mendel's discovery of segregation constituted a relatively revolutionary achievement. (Brannigan 1979, 424)

According to Brannigan, Mendel's work on *Pisum* played a dual role in the history of science. On the one hand, it was well within the paradigm of hybridism.[3] The main problem that Mendel aimed to solve is "the role of hybridization in the evolutionary history of organic forms", which was an unsolved puzzle left by early hybridists. On the other hand, Mendel's work was adopted by the "rediscoverers" to constitute a scientific revolution in the history of the science of heredity. Thus, from his contemporaries' point of view, Mendel's work was part of the paradigm of hybridism, while for the "rediscoverers", some of Mendel's work played a key role in a revolution in the study of heredity in 1900.

Staffan Müller-Wille and Vitězslav Orel (2007) recently provide a similar Kuhnian interpretation.

> Mendel's achievement was a product of normal science, and yet a revolutionary step forward. (Müller-Wille and Orel 2007, 171)

Along with Brannigan, Müller-Wille and Orel argue that Mendel's work was an extension of early hybridists' (especially Gärtner's) work. They convincingly show that there is a continuity throughout the works by Linnaeus, Kölreuter, and Gärtner, linked with anomalies and puzzles under an accepted paradigm. However, there is one substantial difference between Brannigan's and Müller-Wille and Orel's interpretations. Müller-Wille and Orel contend that Mendel's work itself was revolutionary.

From a historical point of view, both accounts are more sophisticated than the theory-based ones discussed in the previous section. Unlike the theory-driven analyses, Mendel's theory is not simply construed as a theory of heredity without argument. Mendel's work is examined in its historical context. For example, the hybridist influence on Mendel is well characterised and highlighted in these Kuhnian

[3]For more discussion on Mendel and hybridism, see Sect. 2.2.

accounts. However, the role of Mendel's work in 1900 has not been articulated adequately. Brannigan is unclear on how and to what extent Mendel's work was adopted and constituted "the Mendelian revolution" in 1900, while Müller-Wille's and Orel's philosophical interpretation that Mendel's work was revolutionary is based on their controversial historical argument that Mendel's work can be understood as a work of heredity.

Nevertheless, these Kuhnian analyses suggest a promising way to analyse and characterise the origin of genetics from Mendel to Bateson: Mendel's work should be better understood as the introduction of an exemplar rather than the proposal of a theory. In order to explore a more sophisticated exemplar-based approach, I find it necessary to examine the conception of exemplar at first.

5.4 A New Interpretation of Exemplar and the Exemplar-Based Approach

Kuhn (1974) defines exemplars as problem-solutions. For Kuhn, exemplars as problem-solutions play an indispensable role in scientific practice. It is Kuhn's novel contribution to introduce the significance of puzzle-solving in scientific practice into the philosophy of science community. However, Kuhn's concept of exemplar still lacks a fuller articulation. In other words, Kuhn's own definition of exemplar is too thin and premature. Many significant problems concerning exemplars are yet to be explored. Firstly, Kuhn says little on how an exemplar is first established or further constructed. As Thomas Nickles (2012, 120) asks, "Where do [exemplars] initially come from?"[4] Although Kuhn is famous for his rejection of the sharp distinction between the context of discovery and of justification and accusing philosophers of ignoring the "temporal development of a theory", Kuhn does not provide a sophisticated account of the construction and temporal development of an exemplar. Secondly, in his elaboration, Kuhn's exemplar is simply illustrated by the examples in the textbooks, lectures, and laboratory exercises. These examples are helpful to provide a rough idea of the application of the exemplars. However, the constituents of an exemplar are never explicitly explicated. Nor is a historical example of an exemplar articulated in an explicit way.[5] Thirdly, Kuhn's exemplar (as a problem-solution) implicitly assumes some pre-existing problems. But where are these pre-existing problems from? Although Kuhn (1970, 103) contends that the shift of accepted exemplars in a scientific revolution necessitates the redefinition of

[4]However, Nickles (2012) still pays insufficient attention to articulate the process of the construction of an exemplar.

[5]There are a few attempts to employ the notion of exemplar to analyse some history cases. For example, Darden (1991) analyses the explanatory virtue of the hybrid crossing in terms of exemplar, while Jeffery Skopek (2011) explores the pedagogical virtue of Mendel's work on peas in terms of exemplar. Unfortunately, the exemplars, for both Darden and Skopek, are simply construed as the examples in the textbook.

research problems, he says little on how the research problems are defined or redefined. Nor is it clear whether problem-defining is a task involving the construction of an exemplar. The significance of problem-defining seems not to be fully recognised by Kuhn. Fourthly, Kuhn fails to explore the characteristics of a good exemplar.[6] It is unclear what makes some exemplars successfully accepted, while others neglected or abandoned.[7] Fifthly, Kuhn is implicit on how an exemplar is or the constituents of an exemplar are used or applied to guide the subsequent research, including constructing a new exemplar, proposing new research problems, solving other problems.

To sum up, Kuhn's definition of exemplar is not well articulated mainly in five ways:

1) The construction of an exemplar is unclear;
2) The constituents of an exemplar are unclear;
3) No detailed historical example of an exemplar is illustrated;
4) What makes an exemplar successfully received is unclear;
5) How an exemplar guides the subsequent research is not explicitly analysed.

Correspondingly, a good reinterpretation of exemplar has to

1') explicate how an exemplar is constructed;
2') identify the constituents of an exemplar;
3') be instantiated by a detailed historical case-study;
4') explore the characteristics of a successfully accepted exemplar;
5') explain how the exemplar can guide subsequent research.

In other words, I not only have to articulate what an exemplar is, what the components of an exemplar are, but also to explore how an exemplar is constructed, how an episode of the history of science can be characterised in terms of exemplars. Moreover, I shall discuss what the characteristics of a good exemplar are, which make it successfully accepted, and examine the instructive function of an exemplar.

One advantage of a Kuhnian exemplar-based account of the history of science is that the significance of the problem-solution in the history and practice of science is well articulated and highlighted. Many of the scientific practices in history are oriented or inspired by some past successful problem-solutions. Kuhn's account of puzzle-solving does capture some essential features of many, though not all, scientific practices in history. Therefore, I would reserve Kuhn's idea that a key constituent of an exemplar is a problem-solution. Furthermore, I argue that an exemplar as the fundamental unit shared by a scientific community should be more than a problem-solution. A well-defined problem itself is at least as important as its solution

[6]Nickles (2012, 128) asks a similar question: "What makes something an exemplar, a problem-cum-solution of the sort that is selected for inclusion in a textbook, widely cited by experts in the field, or the design of an instrument or a technique?"

[7]As I have mentioned, Kuhn (1977) lists five main characteristics of a good theory (or a paradigm). However, the theory (or the paradigm) in this context refers obviously to a disciplinary matrix rather than an exemplar.

in scientific practice. It also has as many normative functions as its solution does. In the history of the sciences, new research problems usually play a vital role to guide the further practice, so the introduction of the new research problem itself is a great scientific achievement. For instance, in *On the Origin of Species*, Charles Darwin introduced many new research problems, which were never thought of or formulated before, like "How will the struggle for existence . . . act in regard to variation? Can the principle of selection, which we have seen is so potent in the hands of man, apply in nature?" (Darwin 1859, 80) In addition, puzzle-solving and problem-defining are two intertwined activities. As I shall argue, an exemplary practice involves the mutually related activities of puzzle-solving and problem-defining. Moreover, it should be noted that problem-defining is much more than proposing a problem. In fact, it usually consists of activities of problem-proposing (i.e. propose an initial problem), problem-refining (i.e. refine an initial problem), and problem-specification (i.e. make an initial problem into some more conceptually specific and experimentally testable problems). The well-defined research problems should be an essential constituent of an exemplar. Thus, my definition of exemplar is as follows:

An exemplar is a set of contextually well-defined research problems and the corresponding solutions.

Firstly, I take an exemplar as *a set of* contextually well-defined research problems and their solutions rather than a *single* problem and its solution. The reason is that *a set of* contextually well-defined problems and their solutions can better reflect the complex aspects of an exemplar as scientific achievement. For example, we may argue that the Morgan school's research on *Drosophila* introduced the problem of the patterns of inheritance of *Drosophila* and its solution. However, in a finer-grained analysis, the Morgan school's research on *Drosophila* introduced *a set of* well-defined research problems (e.g. what is the expected distribution of phenotypes in a certain generation? What is the probability that a particular phenotype will result from a certain mating? What is the frequency of crossing over between two given loci in the chromosomes?) and their solutions.

Secondly, the reason why I define an exemplar as a set of "contextually" well-defined research problems *and* the corresponding successful solutions is that these research problems can only be well-defined and understood in the context of their solutions. In the process of constructing an exemplar, problem-defining and solution-searching are not two independent activities. Rather these are two intertwined activities. On the one hand, problem-solving is obviously dependent on research problems. On the other hand, research problems can be redefined with the process of problem-solving such as conceptualisation and hypothesisation. What is more, as Nicholas Rescher (1984) shows, solutions of problems often generate or propagate new problems.

Thirdly, an exemplar should not be understood in a purely theoretical sense. No exemplar can be constructed in an armchair. Any exemplar must have some non-theoretical components.[8]

A naïve version of the exemplar-based approach can be formulated accordingly as follows.

One should first analyse the history and practice of a science by identifying the research problems. Then, one needs to analyse the solutions and practical efforts to seek solutions, and then provide details about how they were applied to solve the problems.

It is obvious that such an exemplar-based approach is still too vague to be helpful or instructive in analysing the history and practice of science. Thus, I should articulate the components of the solutions of an exemplar in greater detail. However, it should be noted that I do not think that the constituents of the solutions of an exemplar can be characterised in a monistic way. Scientists solve the problems in different ways, so it would be unwise for anyone to try to summarise some universally fundamental parts in their solutions. Therefore, what I would provide is rather a common recipe of an exemplar rather than a definition. By "a common recipe" I mean that an exemplar usually, but not exclusively, consists in such and such components. Here is my common recipe.

An exemplar has five main components: a vocabulary, which is a set of the concepts employed in the problems and solutions; a set of well-defined research problems; a set of practical guides, which specify all the procedures and methodology as means to solve the problems; a set of hypotheses or models, which are proposed to solve the problems; and a set of patterns of reasoning, which indicate how to use other components to solve the problems.[9]

Three points have to be added here. Firstly, these five components are intertwined. For instance, the hypotheses are often formulated on the basis of the results of the experiments by employing the concepts in the vocabulary; the experiments are usually designed and undertaken with the purpose of solving the research problems (e.g. by testing the hypotheses); the concepts in the vocabulary are understood with the help of undertaking the experiments and applying the hypotheses, and so on. Secondly, the vocabulary of an exemplar does not suggest that all the concepts in the vocabulary are first introduced by the exemplar. It is not unusual that in the vocabulary of an exemplar there are some pre-defined concepts. Thirdly, the hypotheses in the exemplar should not be narrowly construed as statements or propositions. Rather I refer "hypotheses" to all kinds of theoretical constructions made by scientists. In history, scientists use different terms to name this kind of work

[8]The concept of exemplar is certainly applicable to many disciplines, including logic and mathematics. But the one discussed in this chapter is only applicable to the empirical sciences.

[9]An example of the patterns of reasoning is the hypothetico-deductive (H-D) model of confirmation, which applies an H-D model of logic to confirm a hypothesis by designing and undertaking the experiments.

like "hypotheses", "assumptions", "principles", "laws", "theories", "models", and "mechanisms".

Thus, correspondingly, the construction of an exemplary practice is a series of intertwined practices of experimentation, problem-defining, conceptualisation, hypothesisation, and reasoning. Experimentation is the practice of designing and undertaking the experiments. Problem-defining is the practice of defining and redefining the research problems. Conceptualisation is the practice of introducing and using a conceptual scheme. Hypothesisation is the practice of theoretical construction to make an explanatory and predictive machinery.[10] Again, all these practices are intertwined and cannot be understood as the independent activities of an exemplary practice.

Therefore, a common recipe for the exemplar-based approach can be summarised as follows.

In order to analyse the history of the practice of a scientific school,[11] *we first should identify the initial problem as the starting point of the research,*[12] *and then trace the way of solving the initial problem by identifying the actual problems to be investigated and the way they occur in the practice, and analysing the process of problem-defining, conceptualisation, experimentation, hypothesisation, and reasoning involved. Then, we should detail the development of the intertwined practices in history to explore the development of a school of scientific practice.*

Before completing my articulation of exemplar and the exemplar-based approach, I highlight one more problem, namely the problem of the reception of an exemplary practice. Why are some exemplary practices successfully received, while others totally neglected or abandoned after some acceptance in the short term? What makes some exemplary practices so successfully accepted? What are the characteristics shared by those successfully accepted exemplary practices?

It is obvious that philosophy alone cannot provide the complete and comprehensive answers to these questions. Why and when an exemplary practice is recognised

[10]Note that I have to emphasise here that there is no universal account of theoretical construction. We have to delve into the historical context to study the process of hypothesisation in any particular case. For example, some hypothesisaitons are better characterised as modelling, while others are better as the discovery of mechanism.

[11]I regard a scientific school as a research community, which is similar to Kuhn's paradigm (1970), Lakatos' research programme (1978), Laudan's research tradition (1977), and Chang's system of practice (2012). Ptolemaic astronomy, Newtonian mechanics, and Mendelian genetics are good examples of scientific schools.

[12]Although I have emphasised that one of the most important contributions of an exemplary practice is the definition of research problems, it is unlikely for a scientist to begin his studies without an initial problem, which was a well-defined research problem. These initial problems might not be interesting at all for the subsequent development of the studies. A clear example is that the initial problem that inspired Morgan to conduct experiments on *Drosophila* was in search for an experimental approach to evolution, but he finally made a great achievement on solving the problems of *Drosophila*'s heredity. Also, it is not unusual that an initial problem is re-formulated in new conceptual frameworks.

and well received by a community of scientists is complex and messy, both sociohistoriographically and philosophically. My interest here is not to attempt to look for universal and comprehensive answers to these questions. Rather, I aim to identify some intellectual characteristics shared by all well received exemplary practices in the history of science, if there are any. I propose that all successfully accepted exemplars share (at least) an essential "intellectual" characteristic: usefulness. A successfully accepted exemplary practice must be useful in the sense that some concepts in the vocabulary, some hypotheses, some research problems, some practical guides, or some patterns of reasoning of the exemplary practice can be used as tools to solve other existing problems or establish new exemplary practices.[13] As Nickles (2012, 128) indicates, "An exemplar candidate, like any tool, will gain status if it shows itself useful in a variety of related situations." It should be noted that usefulness is a necessary condition rather than a sufficient condition of a successfully received exemplary practice.

In summary, I have defined an exemplar as a set of contextually well-defined research problems and the corresponding solutions, which consists of a vocabulary, a set of well-defined research problems, a set of practical guides, a set of hypotheses or models, and a set of patterns of reasoning. I have also proposed that the development of a school of scientific practice can be analysed by identifying the initial research problem as the starting point of the research, and by articulating the way of solving the initial problem with the identification of the actual problems to be investigated and the way they occur in the practice, and by analysing the process of problem-defining, conceptualisation, experimentation, hypothesisation, and reasoning involved. In addition, I have proposed that usefulness is an essential characteristic of a good exemplary practice. In the next section, I shall illustrate how this exemplar-based approach could be applied to analyse the early development in the history of genetics.

5.5 An Exemplar-Based Account of the Origin of Genetics

In his paper, Mendel (1866, 3) was very explicit on his purpose of the study of *Pisum*: to study the development of hybrids in their progeny. More specifically, the initial research problem for Mendel was:

MP1. How could one "determine the number of different forms in which hybrid progeny appear, permit classification of these forms in each generation with certainty, and ascertain their numerical interrelationship"? (Mendel 1866, 4)

In order to make MP1 more experimentally testable and conceptually more specific problems for the further investigation, Mendel reformulated MP1 to a more specific sub-problem:

[13] For the full articulation of "usefulness", see Chap. 6.

MP2. What are the changes for each pair of differing traits [i.e. dominant/recessive traits], selected in the pre-experiment practice, in the offspring of *Pisum*? Or, what is the law deducible from the changes for each pair of differing traits selected in the successive generations? (Mendel 1866, 7)

With contextually intertwined activities by problem-defining, conceptualisation, experimentation, hypothesisation, and reasoning, Mendel established an exemplary practice on the study of hybrid development, summarised in Table 5.4.

The significance of Mendel's exemplary practice was to some extent overlooked in the study of heredity until it was adopted by de Vries in his study of pangenesis. De Vries's initial problem (DP1) was to experimentally test the principle (DH1) that the specific characters of organisms are composed of distinct units (de Vries 1900a, 1900b). It should be noted that DH1 was a reformulated version of the hypothesis (DH1') in the theory of pangenesis (1889) that every hereditary characteristic has its special kind of pangens. De Vries had struggled to find a way of analysing the data based on his hybridisation experiments until he recognised that Mendel's problem-specification (MP1 → MP2), concepts of dominance and recessiveness, and law of composition of hybrid fertilising cells (MH3) are useful. By incorporating Mendel's problem-defining, conceptualisation, and hypothesisation, de Vries constructed an exemplary practice on the study of pangenesis, which can be summarised as in Table 5.5.

Correns' initial concern was the xenia question (CP1), that is, whether foreign pollen had a direct influence on the characteristics of the fruit and seed. In 1896 he began studying this problem in the case of *Pisum*. After reading Mendel's paper, Correns immediately recognised the usefulness of Mendel's work. The purpose of his 1900 paper was to test Mendel's work on *Pisum*. Thus, CP1 was specialised into another problem.

CP2. Is Mendel's observation and the law on *Pisum* verifiable?

In order to test Mendel's observation and analysis, Correns followed Mendel to focus on a pair of differing traits. (CG1) In other words, a more specific problem occurs.

CP3. Is Mendel's observation and law concerning a pair of differing traits confirmable?

Correns' exemplary practice on testing Mendel's study of *Pisum* can be summarised in Table 5.6.

By analysing Mendel's, de Vries', and Correns' exemplary practices, I argue that there are four constituents of Mendel's exemplary practice preserved (or preserved with minor modifications) and passed on in the successors' exemplary practices, despite their different initial research problems. And all these constituents well account for Mendel's major contributions to the history of genetics, characterised by the historians.

First of all, as some historians (e.g. Olby 1979; Müller-Wille and Orel 2007) have already pointed out, one of Mendel's important achievements was that his approach

Table 5.4 Mendel's exemplary practice on *Pisum*

Research problems	Vocabulary	Practical guides	Hypothesisation	Experiments	Patterns of reasoning
MP1. How could one "determine the number of different forms in which hybrid progeny appear, permit classification of these forms in each generation with certainty, and ascertain their numerical interrelationship"? MP2. What are the changes for each pair of differing traits, selected in the pre-experiment practice, in the offspring of *Pisum*? Or, what is the law deducible from the changes for each pair of differing traits selected in the successive generations? MP3. Is the law of development concerning a pair of traits (i.e. H1) still applicable when several traits are united in the hybrid of *Pisum* through fertilisation? MP4. How can MH1 and MH2 be explained in terms of seed and pollen cells?	Dominating constant trait (A) Recessive constant trait (a) Hybrid trait (aa) Kinds of germinal cell (e.g. A') Kinds of pollen cell (e.g. a')	MG1. The selection of experimental plants (MC1, MC2, MC3) MG2. The selection of morphological traits of peas (C4) MG3. Other pre-experimental procedures (e.g. places of growing) Experimental procedures for various experiments	MH1. "Of the seeds formed by the hybrids with one pair of differing traits, one half again develop the hybrid form while the other half yield plants that remain constant and receive the dominating and the recessive character in equal shares." (The law of development concerning a pair of differing traits) MH1'. In the nth generation the distribution of the dominant constant, hybrid, and recessive constant traits is the 2n-1: 2n: 2n-1 ratio if "on the average, equal fertility for all plants in all generations, and if one considers, furthermore, that half of the seeds that each hybrid produces yield hybrids again while in the other half the two traits become constant in equal proportions."	ME1 ME2 ME3 MEn ME1' ME2' ME1" ME2"	H-D confirmation

			MH2. "The progeny of hybrids in which several essentially different traits are united represent the terms of a combination series in which the series for each pair of differing traits are combined. . . [T]he behaviour of each pair of differing traits in a hybrid association is independent of all other differences in the two parental plants." (The law of combination of differing traits) MH2'. If n designates the number of pairs of differing traits in the parental plants, then 3n is the number of different trait combination, 4n is the number of, 2n is the number of combinations that remain constant. MH3. Pea hybrids form germinal and pollen cells that in their composition correspond in equal numbers to all the constant forms resulting from the combination of traits united through fertilisation.		

Table 5.5 De Vries' exemplary practice on *Character-Unit*

Research problems	Vocabulary	Practical guides	Hypotheses	Experiments	Patterns of reasoning
DP1. How to experimentally test the principle DH1? DP2. What is the change of a pair of antagonistic traits of hybrid?	Dominating trait (A) Recessive trait (a) Dominating characteristic Recessive characteristic Units	DG1	DH1. The specific characters of organisms are composed of distinct units. DH2. In the hybrids two antagonistic characteristics lie next to each other. DH2'. In the hybrids the pangens carrying two antagonistic characteristics lie next to each. DH3. In vegetative life only the dominating characteristics is visible. DH4. In the formation of pollen grains and ovules these characteristics separate and behave independently. DH4'. In the formation of pollen grains and ovules these pangens separate. DH5. The pollen grains and ovules of monohybrid have the pure characteristic one of the parents. DH6. $(d + r)(d + r) = d^2 + 2dr + r^2$ DH7. $(d + r)d = d^2 + dr$ DH7'. The offspring of the hybrid seeds with the pollens of one of the two parents have two combinations of pangens: One is with two pangens both carrying the same one parental characteristic, while the other is with two pangens carrying two parental characteristics each. DH7''. Half of the offspring of the hybrid seeds with the pollens of one of the two parents have one of the two parental traits, while the other half have the hybrid trait (i.e. the dominating trait).	DE1 DE2	H-D confirmation

Table 5.6 Correns' exemplary practice on *Mendel's Study of Pisum*

Research problems	Vocabulary	Practical guides	Hypotheses	Experiments	Patterns of reasoning
CP1. Does foreign pollen have a direct influence on the characteristics of the fruit and seed? CP2. Is Mendel's observation and the law on *Pisum* verifiable? CP3. Is Mendel's observation and law concerning a pair of differing traits confirmable? CP4. Is Mendel's observation and law concerning two or more pair of differing traits confirmable? CP4'. Is Mendel's LCD confirmable? CP5. Is Mendel's LCC universally applicable?	Anlage Dominating trait (A) Recessive trait (a) Dominating anlage Recessive anlage	CG1	CH1. In the fusion of the reproductive nuclei, the anlage for the recessive trait is suppressed by the one for the dominating trait. Prior to the definitive formation of the reproductive nuclei a complete separation of the two anlagen occurs, so that one half of the reproductive nuclei receive the anlage for the recessive trait, the other half the anlage for the dominating trait. CH2. In the hybrid, reproductive cells are produced in which the anlagen for the individual parental characteristics are contained in all possible combinations, but both anlagen for the same pair of traits are never combination. Each combination occurs with approximately the same frequency.	CE1 CE2	H-D confirmation

to the study of the problem of development by focusing on the paired traits in the successive generations. Mendel's important observations and hypotheses were all about paired traits of hybrids and their progeny. This can be clearly interpreted as the problem-specification (MP1 → MP2). As Müller-Wille and Orel (2007, 211) indicate, "Mendel's focus on character pairs was not only an important methodological step, but had immediate consequences for his theorizing." The significance of Mendel's problem-specification (MP1 → MP2) is also reflected in its reception at the beginning of the twentieth century. Although de Vries, Correns, Tschermak, and Bateson were not studying hybrid development, all of them adopted Mendel's approach to concentrate on paired traits. It is no surprise that de Vries' problem-specification (DP1 → DP2) was indebted to Mendel's problem-specification (MP1 → MP2).

> In every crossing experiment only a single character or a definite number of them is to be taken into consideration . . . for experimental purposes the simplest conditions are presented by hybrids whose parents differ from each other in one trait only. (de Vries 1966, 108)

Tschermak (1900a) also adopted the Mendelian problem-specification (MP1 → MP2) to study his research problem. In particular, Bateson was explicit on the point that a significant lesson learnt from Mendel in the study of heredity was that of focusing on differing traits (MP2).

> [T]he subjects of experiment should be chosen in such a way as to bring the laws of heredity to a real test. For this purpose the first essential is that the differentiating characters should be few, and that all avoidable complications should be got rid of. Each experiment should be reduced to its simplest possible limits . . . [I]t is certain that by similar treatment our knowledge of heredity may be rapidly extended. (Bateson 1902, 16)

Thus, I argue that what de Vries, Correns, Tschermak, and Bateson in fact learnt from Mendel here is a way of refining a general problem into a more specific one. Despite beginning with different initial research problems, de Vries, Correns, and Bateson, influenced by Mendel's work, all found that refining their initial problems into a better defined and more narrowly scoped problem on paired traits was helpful in the further investigation.

Another contribution of Mendel's work was his exemplary use of the terms "dominant" and "recessive" to conceptualise the paired traits and analyse the statistical relation of them. Though the phenomenon of dominance had been observed by many (e.g. Knight 1799; Goss 1824; Seton 1824) by the first half of the nineteenth century, Mendel was the first to conceptualise the phenomenon in terms of dominance/recessiveness, and record and analyse the statistical relation of dominant and recessive traits. Mendel's terminology was important for his work in the sense that it lay down the conceptual foundation for his problem-specification (MP1 → MP2), analysis of data, recognition of the statistical regularity (e.g. the 3: 1 ratio) and proposal of the hypotheses. It should be highlighted that the significance of Mendel's terminology and his statistical analysis were intertwined. The statistical analysis could not be made without the concepts of "dominant" and "recessive", while the introduction of the concepts of dominance and recessiveness was not so interesting if no statistical regularity was obtained.

Mendel's terminology also enlightened the study of heredity around 1900. The terms "dominant" and "recessive" were adopted by de Vries, Correns, Tschermak, and Bateson, though their usages were different from Mendel's in some respects (for a summary, see Table 5.4). Correspondingly, the statistical analysis of the dominating and recessive traits was also introduced into the study of heredity, especially by de Vries (1900a, 1900b, 1900c) and Bateson (1902).[14]

I have to emphasise that the concepts of dominance and recessiveness were important in the history of genetics because they were useful in conceptualisation, hypothesisation, and idealisation rather than because they were essential conceptual components, which were invariantly shared by both Mendel and the followers.

Thirdly, Mendel's law of composition of hybrid fertilising cells (MH3) was also particularly exemplary. What was really novel in Mendel's MH3 is the correspondence of the statistical relations of morphological traits and of germinal and pollen cells.[15] The hypothesis concerning the morphological-cellular correspondence proposed by Mendel became a key to advance the study of heredity three decades later. The biggest difficulty identified by Bateson (1902) in the study of heredity at the turn of the twentieth century was the lack of a reliable approach to studying the physical basis of heredity. In fact, there were a few theories of heredity concerning the physical basis. Weismann's theory of germ-plasm (1892) and de Vries' theory of pangenesis (1889) were two representative ones. However, neither provided a feasible way to test the theories. In other words, the relation of visible characters and invisible "physical basis of heredity" was untestable experimentally. The state of the study of heredity around 1900 was such that, as Bateson (1902, 3) neatly summarised, "[n]o one has yet any suggestion, working hypothesis, or mental picture that has thus far helped in the slightest degree to penetrate beyond what we see." In 1900, de Vries adopted and revised Mendel's hypothesis on the morphological-cellular correspondence to support his theory of pangenesis. In addition, de Vries also limited the application of his trait-characteristics correspondence in the case of true hybrids. Correns made "a significant step beyond Mendel" by reformulating Mendel's hypothesis as the trait-anlage correspondence (CH1, CH2), though he was not very clear on the implication of this reformulation in the study of heredity. Bateson was the first to make a sophisticated attempt to incorporate Mendel's morphological-cellular correspondence into the study of heredity. Bateson revised Mendel's hypothesis as the trait-paired allele determination. It is determination rather than correspondence, because Bateson explicitly talked of "the bearers of the character". In addition, in contrast to the limited applicability of Mendel's correspondence, Bateson's trait-paired alleles determination was applicable broadly in certain phenomena of alternative inheritance. Though Mendel's hypothesis concerning the morphological-cellular correspondence was not adopted without modification in de Vries', Correns', and Bateson's work, it was really helpful to set out an approach to working on the "inward and essential nature" of heredity.

Fourthly, Mendel's other contribution to the study of heredity in 1900 was his exemplary mathematical approach (Olby 1997). Mendel made a novel statistical

[14]For an in-depth analysis of the historical development of the concept of dominance, see Chap. 8.

[15]It is worth noting that correspondence is a weaker notion than determination. Mendel never used the notion of determination, or causation in MH3.

analysis of the developmental series.[16] He denotes the dominant (constant) trait A, the hybrid trait Aa, the recessive (constant) trait a, and the distribution of these traits in the F_2 generation (A + 2Aa + a). All these symbolic notations are much more important and useful than at first appears. All Mendel's three laws can be formulated in the equations in terms of these notations.

MH1: $A + 2Aa + a$

MH2: $(A + 2Aa + a)\ (B + 2Bb + b) = AB + Ab + aB + 2ABb + 2aBb + 2AaB + 2AaB + 4AaBb$

MH3: $\frac{A'}{A'} + \frac{A'}{a'} + \frac{A'}{a'} + \frac{a'}{a'} = A + 2Aa + a$

Moreover, Mendel's MH1 and MH2 were introduced and articulated with the help of these notations and the mathematical manipulation of these. Mendel's mathematical formulation was also adopted and further developed by de Vries. The distribution of characteristics in the F_2 generation was formulated by de Vries as $d^2 + 2dr + r^2$. The move from A + 2Aa + a to $d^2 + 2dr + r^2$ was a breakthrough in the history of genetics. The equation $(d + r)\ (d + r) = d^2 + 2dr + r^2$ implicitly suggested that the phenomenon of the separation of hereditary characteristics took place within pollen grains and ovules. This lays down the cornerstone for a later conception of particulate inheritance. Mendel's laws, and the concepts of allelomorph (Bateson 1902, 1909), factor (Punnett 1905; Morgan et al. 1915), and gene (Morgan 1926) were all articulated with the help of similar notations. Although, as many (e.g. Olby 1979) have pointed out, Mendel himself never had the conception of pairs of hereditary elements determining the morphological trait, his mathematical approach still played an indispensable role in the founding of genetics as a school of scientific practice. Therefore, as I have shown, Mendel's four significant contributions, namely, the focus on a pair of differing traits, the conceptions of dominance and recessiveness and their statistical relation, the proposal of the hypothesis concerning the morphological-cellular correspondence, and the mathematical approach, can be well accounted as problem-specification, conceptualisation, and hypothesisation.

Thus, I argue that Mendel's work is well characterised as an exemplary practice of the development of pea hybrids in their progeny. More specifically, *Mendel introduced a set of contextually well-defined research problems on the development of hybrids in their progeny and the corresponding solutions, and some components of his exemplary practice greatly inspired and influenced de Vries', Correns', Tschermak', and Bateson's work, and lay down the cornerstone for the study of heredity in the twentieth century*. In particular, as I have argued earlier in this section, Mendel's focus on a pair of differing traits (MP1 → MP2), the proposal of the conceptions of dominance and recessiveness and their statistical relation, the introduction of the hypothesis of the morphological-cellular correspondence (MH3), and his mathematical approach made an enormous impact on de Vries', Correns' and

[16]It should be noted that as I have mentioned in Chap. 2, Mendel was not the first to apply the mathematical/statistical approach to biological studies.

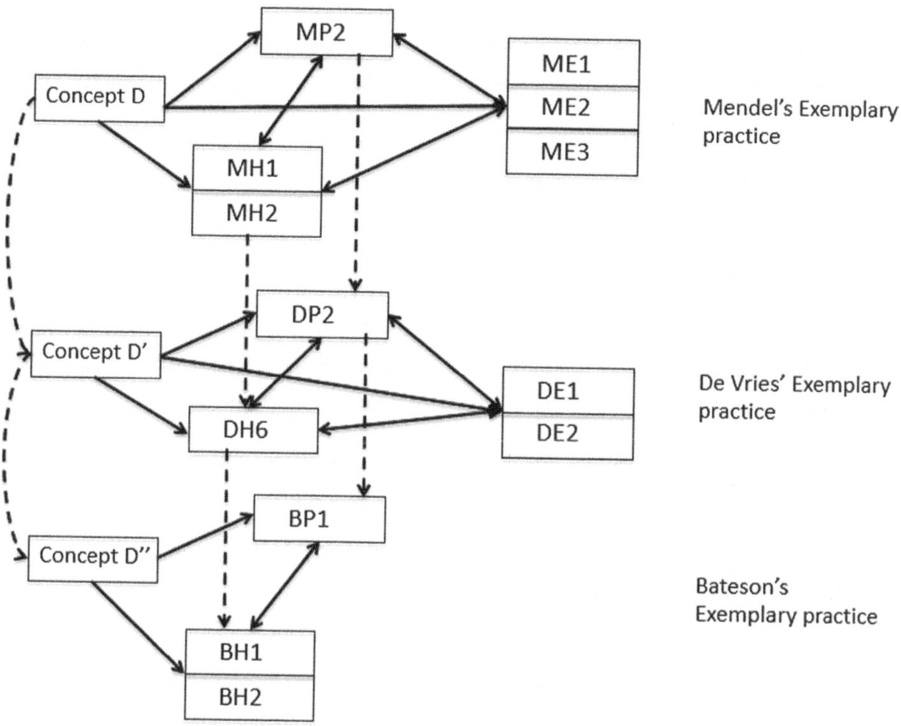

Fig. 5.1 Partial Representation of the Chain of the Exemplary Practices from Mendel to Bateson (For a detailed elaboration, see Sect. 8.5)

Bateson's' work on heredity. Similarly, *de Vries', Correns', and Bateson's works can also be characterised as different exemplary practices, which also inspired and influenced the successors' work* (e.g. Castle and Allen 1903; Castle 1903; Punnett 1905; Raynor and Doncaster 1905; Hurst 1906) on heredity in the first decade of the twentieth century. Therefore, the origin of genetics from Mendel to Bateson, I argue, can be characterised as a chain of exemplary practices, as illustrated in Fig. 5.1.

In the history of genetics, the earlier exemplary practices were accepted and learnt by the successor practitioners. It should be noted that to say that the practitioners accept an exemplary practice does not mean that all the components of that exemplary practice are accepted and shared dogmatically. Instead what is accepted and shared by all practitioners is the way of defining the problems and of solving these problems. In the case of the origin of genetics, what de Vries, Correns, Tschermak, and Bateson all shared and accepted is Mendel's problem-defining and problem-solving. Nevertheless, they still differed in how to understand the components of Mendel's exemplary practice and how to use them (or some of them) to solve their problems. For the "rediscoverers", Mendel's conceptualisation, hypothesisation, experimentation, and reasoning were just tools to solve Mendel's problems, but some of the tools were useful to their own research problems. Thus, Mendel's vocabulary, hypotheses, practical guides, experiments and patterns of reasoning were accepted as tools to solve Mendel's problem of hybrid development and to solve successors' problems. In addition, to say that the exemplary practices are accepted does not mean that they are accepted invariantly by all the late practitioners. For example, it would be plausible to argue that de Vries and Bateson accepted and worked on Mendel's exemplar practice, while Hurst and Punnett accepted and worked on Bateson's exemplary practice rather than Mendel's. This is why I argue that the origin of genetics is better characterised as a chain of exemplary practices rather than a set of exemplary practices.

To some extent, characterising Mendel's work as the introduction of new problems and their solutions is not a completely new idea. In particular, the significance of Mendel's introduction of new research problems had been highlighted by many historians (e.g. Sandler and Sandler 1985; Bowler 1989). In particular, Iris Sandler and Laurence Sandler (1985, 69) explicitly pointed out that "Mendel . . . defined his problem in purely genetic terms, and produced a correct and amazingly complete answer." However, one crucial difference between Sandlers' and my interpretation is that Sandlers' focus is mainly historical. Little is said about what the problem is and what the answer is, or how the problem and its solution influence the successor's work methodologically, conceptually, and theoretically. And this is what the exemplar-based approach can contribute to examine these issues.

5.6 Conclusion

Still some may challenge the superiority of the exemplar-based approach to the theory-driven one. Even if the exemplar-based approach helps us have a plausible account of the origin of genetics, it just provides an alternative way to characterise the history of genetics. It is still too early to conclude that the exemplar-based approach is a better fit than the theory-driven one in analysing the origin of genetics. In response, I argue that this is unlikely given two main advantages of the exemplar-based analysis.

First, the exemplar-based approach is exempt from any kind of problem of identification. According to the exemplar-based account, there is no need to identify any "essence" of genetics. No single component of exemplary practices is required to be invariantly preserved in a chain of exemplary practices. In contrast, the theory-driven account cannot solve the problem of theory-identification completely. Second, the exemplar-based approach enables us to articulate multi-facets of the origin of genetics. The non-theoretical aspects of the practice and their relation to the theoretical ones have been examined and highlighted in greater detail. Therefore, I argue that the exemplar-based approach not only provides us a new philosophical way to analyse the origin of genetics, but also enables us to have a historiographically more sophisticated account.

In summary, firstly I have argued that the theory-based approach to the origin of genetics is highly problematic. It oversimplifies the theoretical development in the history of genetics and overlooks the significance of the non-theoretical development. Secondly, I have proposed the exemplar-based approached. Thirdly, I have argued that the origin of genetics from Mendel to Bateson should be better characterised as a chain of exemplary practices.[17] I have shown that this exemplar-based analysis of the origin of genetics makes a fuller understanding of the development of genetics in the early period.

References

Bateson, William. 1902. *Mendel's Principles of Heredity: A Defence*. Cambridge: Cambridge University Press.

———. 1909. *Mendel's Principles of Heredity*. Cambridge: Cambridge University Press.

Bowler, Peter J. 1989. *The Mendelian Revolution: The Emergence of Hereditarian Concepts in Modern Science and Society*. London: The Athlone Press.

Brannigan, Augustine. 1979. The Reification of Mendel. *Social Studies of Science* 9 (4): 423–454.

Cartwright, Nancy, Towfic Schomar, and Mauricio Suárez. 1995. The Tool Box of Science. In *Theories and Models in Scientific Progresses*, ed. William E. Herfel, Wladyslaw Krajewski, Ilkka Niiniluoto, and Ryszard Wojcicki, 137–149. Amsterdam: Rodopi.

Castle, W.E. 1903. Mendel's Law of Heredity. *Science* 18 (456): 396–406.

Castle, W.E., and Glover M. Allen. 1903. The Heredity of Albinism. *Proceedings of the American Academy of Arts and Sciences* 38 (21): 603–622.

Chang, Hasok. 2012. *Is Water H_2O? Evidence, Realism and Pluralism*. Dordrecht: Springer.

Correns, Carl. 1900. G. Mendels Regel über das Verhalten der Nachkommenschaft der Rassenbastarde. *Berichte der Deutschen Botanischen Gesellschaft* 18 (4): 158–168.

Darden, Lindley. 1991. *Theory Change in Science: Strategies from Mendelian Genetics*. Oxford: Oxford University Press.

Darwin, Charles. 1859. *On the Origin of Species*. London: John Murray.

Davenport, Charles B. 1901. Mendel's Law of Dichotomy in Hybrids. *The Biological Bulletin* 2 (6): 307–310.

Goss, John. 1824. On the Variation in the Colour of Peas, Occasioned by Cross Impregnation. *Transactions of the Horticultural Society of London* 5: 234–236.

Hacking, Ian. 1983. *Representing and Intervening: Introductory Topics in the Philosophy of Natural Science*. Cambridge: Cambridge University Press.

Hurst, C.C. 1906. On the Inheritance of Coat Colour in Horses. *Proceedings of the Royal Society of London, Series B* 77 (519): 388–394.

Kitcher, Philip. 1984. 1953 and All That: A Tale of Two Sciences. *The Philosophical Review* 93 (3): 335–373.

———. 1989. Explanatory Unification and the Causal Structure of the World. In *Scientific Explanation*, ed. Philip Kitcher and Wesley C. Salmon, 410–505. Minneapolis, MN: University of Minnesota Press.

Knight, Thomas Andrew. 1799. An Account of Some Experiments on the Fecundation of Vegetables. *Philosocial Transactions of the Royal Society of London* 89: 195–204.

[17] I have to emphasise that the origin of genetics from Mendel to Bateson discussed in this paper is definitely not a complete and comprehensive picture of the origin of genetics. As Olby (1985) has suggested, there are multiple origins of genetics. What I focus on here is only one path to genetics. More precisely speaking, my task is to explore a new exemplar-based way to analyse and understand the development from Mendel's (1866), de Vries' (1900a, 1900b, 1900c), Correns' (1900), Tschermak's (1900a, 1900b) to Bateson's work (1902).

Kuhn, Thomas Samuel. 1970. *The Structure of Scientific Revolutions*. 2nd ed. Chicago, IL: The University of Chicago Press.

———. 1974. Second Thoughts on Paradigms. In *The Structure of Scientific Theories*, edited by Frederick Suppe, 1st ed., 459–482. Urbana, IL: The University of Illinois Press.

———. 1977. Objectivity, Value Judgment, and Theory Choice. In *The Essential Tension: Selected Studies in Scientific Tradition and Change*, 320–339. Chicago, IL: The University of Chicago Press.

Lakatos, Imre. 1978. Falsification and the Methodology of Scientific Research Programmes. In *The Methodology of Scientific Research Programme*, edited by John Worrall and Gregory Currie, 8–101. Cambridge: Cambridge University Press.

Laudan, Larry. 1977. *Progress and its Problems*. Berkeley and Los Angeles: The University of California Press.

Mendel, Gregor. 1866. Versuche über Pflanzenhybriden. *Verhandlungen des Naturforschenden Vereins Brünn* IV (1865) (Abhandlungen): 3–47.

Morgan, Thomas Hunt. 1926. *The Theory of the Gene*. New Haven, CT: Yale University Press.

Morgan, Thomas Hunt, Alfred Henry Sturtevant, Hermann Joseph Muller, and Calvin Blackman Bridges. 1915. *The Mechanism of Mendelian Heredity*. New York: Henry Holt and Company.

Moss, Lenny. 2003. *What Genes Can't Do*. Cambridge, MA: The MIT Press.

Müller-Wille, Staffan, and Vitězslav Orel. 2007. From Linnaean Species to Mendelian Factors: Elements of Hybridism, 1751-1870. *Annals of Science* 64 (2): 171–215.

Nickles, Thomas. 2012. Some Puzzles about Kuhn's Exemplars. In *Kuhn's The Structure of Scientific Revolutions*, ed. Vasso Kindi and Theodore Arabatzis, 112–133. London and New York: Routledge.

Olby, Robert Cecil. 1979. Mendel No Mendelian? *History of Science* 17 (1): 53–72.

———. 1985. *Origins of Mendelism*. 2nd ed. Chicago, IL: The University of Chicago Press.

———. 1997. Mendel, Mendelism and Genetics. *MendelWeb*. 1997. http://www.mendelweb.org/MWolby.html.

Punnett, Reginald Crundall. 1905. *Mendelism*. London: Macmillan and Co., Limited.

Raynor, G.H., and L. Doncaster. 1905. Experiments on Heredity and Sex-Determination in Abraxas Grossulariata. In *In Report of the 74th Meeting of the British Association of the Adavancement of Science, Cambridge 1904*, 594–595. London: John Murray.

Rescher, Nicholas. 1984. *The Limit of Science*. Berkeley, CA: The University of California Press.

Ruse, Michael. 1973. *The Philosophy of Biology*. London: Hutchinson & Co.

Sandler, Iris, and Laurence Sandler. 1985. A Conceptual Ambiguity That Contributed to the Neglect of Mendel's Paper. *History and Philosophy of the Life Sciences* 7 (1): 3–70.

Seton, A. 1824. Note by the Secretary. *Transactions of the Horticultural Society of London* 5: 236–237.

Simunek, Michal, Uwe Hoßfeld, and Olaf Breidbach. 2011. 'Rediscovery Revised' – The Coopertation of Erich and Armin von Tschermak-Seysenegg in the Context of the 'Rediscovery' of Mendel's Law in 1899–1901. *Plant Biology* 13 (6): 835–841.

———. 2012. 'Further Development' of Mendel's Legacy? Erich von Tschermak-Seysenegg in the Context of Mendelian-Biometry Controversy, 1901-1906. *Theory in Biosciences* 131 (4): 243–252.

Skopek, Jeffrey M. 2011. Principles, Exemplars, and Uses of History in Early 20th Century Genetics. *Studies in History and Philosophy of Biological and Biomedical Sciences* 42 (2): 210–225.

von Tschermak, Erich. 1900a. Über Künstliche Kreuzung bei Pisum Sativum. *Berichte der Deutschen Botanischen Gesellschaft* 18 (6): 232–239.

———. 1900b. Über Künstliche Kreuzung bei Pisum Sativum. *Zeitschrift für das Landwirtschaftliche Versuchswesen in Oesterreich* 3: 465–555.

de Vries, Hugo. 1889. *Intracellulare Pangenesis*. Jean: Gustav Fischer.

———. 1900a. Das Spaltungsgesetz der Bastarde (Vorlaufige Mittheilung). *Berichte der Deutschen Botanischen Gesellschaft* 18 (3): 83–90.

———. 1900b. Sur la Loi de Disjonction des Hybrides. *Comptes Rendus de l'Academie des Sciences (Paris)* 130: 845–847.

———. 1900c. Sur les Unités des Caractères Spécifiques et Leur Application à l'étude des Hybrides. *Revue Générate de Botanique* 12: 257–271.

———. 1966. The Law of Segregation of Hybrids. In *The Origin of Genetics: A Mendel Source Book*, edited by Curt Stern and Eva R. Sherwood, translated by Evelyn Stern, 107–17. San Francisco, CA: W. H. Freeman and Company.

Waters, C. Kenneth. 2004. What was Classical Genetics? *Studies in History and Philosophy of Science* 35 (4): 783–809.

Weismann, August. 1892. *Das Keimplasma: Eine Theorie Der Vererbung*. Jena: Gustav Fischer.

Weldon, Walter Frank Rapheal. 1902. Mendel's Laws of Alternative Inheritance in Peas. *Biometrika* 1 (2): 228–254.

Chapter 6
A Functional Account of the Progress in Early Genetics

Abstract In this chapter, I argue that the exemplar-based approach motivates a new functional approach to scientific progress, which makes a better account of the progress in the history of genetics. First of all, motivated by the exemplar-based approach, I propose a new functional approach to scientific progress, in which scientific progress is defined in terms of usefulness of problem-defining and problem-solving. Secondly, I further develop a functional account of the progress in early genetics. Thirdly, I argue that the new functional approach well resolves the problems of the traditional functional approach. Fourthly, I highlight the advantages of my new functional account over the epistemic and semantic accounts and dismiss some potential objections to my account.

Keywords Scientific progress · Usefulness · Problem-defining · Problem-solving · Origin of genetics

6.1 Scientific Progress and the Origin of Genetics

Few would deny that genetics has progressed greatly since Charles Darwin's proposal of the theory of pangenesis (1868). However, it is not an easy task to make sense of the progress in the history of genetics. In general, there are three popular approaches to characterising scientific progress: the epistemic approach (e.g. Bird 2007), the semantic approach (e.g. Niiniluoto 2014), and the functional-internalist approach (e.g. Kuhn 1962; Laudan 1977, 1981). The epistemic approach defines progress in terms of knowledge. The semantic approach defines progress in terms of truth or verisimilitude. The functional-internalist approach construes progress in terms of functions of scientific practice. Correspondingly, the epistemic account of the progress in the history of genetics is that the development of genetics is progressive if it shows the accumulation of scientific knowledge. According to the semantic account, the development of genetics is progressive if either it shows the accumulation of true scientific beliefs or it shows the increasing approximation of true scientific beliefs. According to the functional-internalist account, the development of genetics is progressive if it shows

Y. Shan, *Doing Integrated History and Philosophy of Science: A Case Study of the Origin of Genetics*, Boston Studies in the Philosophy and History of Science 320,
https://doi.org/10.1007/978-3-030-50617-9_6

the success of the fulfilment of a certain function (for example, problem-solving), where the fulfilment of the function can be judged by scientists at that time.

Unfortunately, none of these accounts well characterises the progress in early genetics. It is controversial that the knowledge of heredity accumulated from Darwin to William Bateson. Darwin's theory of pangenesis was abandoned by early Mendelians at the beginning of the twentieth century. It is doubtful that the theory of heredity was even approximating the truth in that period. It is not obvious that Bateson's Mendelian principles of heredity (1902) was closer to the truth than Hugo de Vries' law of segregation (1900b). Nor is it clear that Carl Correns' Mendelian rule (1900) solved more problems than Francis Galton's law of ancestral heredity (1889).

This chapter develops and defends a new functional account of the progress in the history of genetics. In Sect. 6.2, motivated by the exemplar-based approach, I propose a new functional approach to scientific progress, in which scientific progress is defined in terms of usefulness of problem-defining and problem-solving. In Sect. 6.3, I further develop a functional account of the progress in early genetics. In Sect. 6.4, I argue that the new functional approach well resolves the problems of the traditional functional approach. In Sect. 6.5, I highlight the advantages of my new functional account over the epistemic and semantic accounts and dismiss some potential objections to my account.

6.2 A New Functional Approach to Scientific Progress

In Chap. 5, I have proposed that usefulness is an essential characteristic of a good exemplary practice. Thus, it seems natural to argue that an exemplary practice exhibits more progress than another if the former is more useful than the latter. Accordingly, I propose this new functional account of scientific progress:

Science progresses if more useful exemplary practices (or, more useful research problems and their corresponding solutions) are proposed.

It should also be noted that more useful exemplary practices could be understood in two ways: (1) *more useful* exemplary practices, and (2) *more* useful exemplary practices. What I take here is (1), the qualitative sense:

An exemplary practice (or, a set of contextually well-defined problems and their solutions) is useful if and only if the way of defining and solving the research problems in that exemplary practice is repeatable, and provides a reliable framework for further investigation to solve the unsolved problems and to generate more testable research problems across more different areas (or disciplines).

This notion of usefulness encompasses four virtues: repeatability, problem-solving success, problem-defining novelty, and interdisciplinarity. The repeatability of the way of defining and solving research problems is a prerequisite for the recognition of its usefulness. For example, one main reason for W. F. R. Weldon (1902a, 1902b) to resist the Mendelian approach to the study of heredity was that he failed to repeat Mendel's way of distinguishing the dominant from recessive characters as a part of the solution to the problem of the transmission of morphological traits. In contrast, Correns' acceptance of the usefulness of the Mendelian approach (1900) was based on his successful repetition of Mendel's exemplary practice, including problem-defining, conceptualisation, hypothesisation, and experimentation. Moreover, the mixed reception of the Mendelian approach in the first decade of the twentieth century was also due to the different results of the application of the Mendelian approach to the problems of the transmission of characters in other species. Hugo de Vries' acceptance of the Mendelian approach (1900a, 1900b) was because it was successfully applied to the study of the transmission of the morphological traits in various plant species (e.g. *Lychnis*, *Papaver*, and *Solanum*), while scepticism arose from the unfavourable results of the application (e.g. Whitman 1904; McCracken 1905, 1906, 1907; Reid 1905; Prout 1907; Saunders 1907; Hart 1909; Holmes and Loomis 1909). The problem-solving success has been widely acknowledged as a virtue of scientific practice (e.g. Kuhn 1970b; Laudan 1977; Nersessian 2008), while the problem-defining novelty is also highlighted in Chap. 5. In addition, the interdisciplinarity is an important virtue in scientific practice. Our science advances with so many interactions of different disciplines, for example, astrophysics, biochemistry, bioinformatics. The interdisciplinarity of an exemplary practice helps to widen the scope and explore the novel ways of scientific inquiry. As I shall show in greater detail in Sect. 6.3, Mendel's problems and solutions introduced in his study on hybrid development were useful in the sense that they were repeatable in practice and provided the foundation for the twentieth century study of heredity to solve the problems of transmission of the morphological traits of other species and to generate more potential testable research problems across the areas like cytology, evolution, and heredity.

In addition, it should be noted that the acceptance of usefulness is relative to scientific communities. A particular set of problem-defining and problem-solving might be taken as useful by some scientific communities but not others. The Mendelian-Biometrician controversy in the first decade of the twentieth century well illustrates this point. The Mendelians, led by Bateson, were optimistic on the future of the Mendelian approach to the study of heredity, while the Biometricians, led by Weldon, doubted the usefulness of Mendel's approach (especially the conceptualisation, hypothesisation, and experimentation) to study the phenomena of heredity. In other words, the Biometricians overlooked the progressive element of the Mendelian approach due to their failure of the recognition of its usefulness. Thus, the usefulness of a certain set of problem-defining and problem-solving may not be obvious to a scientific community. The progress thus achieved is not judged or known by the community in a straightforward way.

6.3 How Early Genetics Progressed

From a close reading of Darwin's, Mendel's, de Vries', Correns', and Bateson's writings, I argue that the progress in early genetic could be well illustrated and explained as a process of the increase of the usefulness of new research problems and their solutions.

The study of heredity in the nineteenth century was galvanised by Darwin's study of evolution, as he was looking for a mechanism of heredity to support his theory of evolution by natural selection (1859). To start, Darwin (1868, 357) introduced a set of the problems of heredity to be solved:

DQ1. How is it possible for a character possessed by some remote ancestor suddenly to reappear in the offspring? (The Problem of Atavism)
DQ2. How can the effects of increased or decreased use of a limb be transmitted to the child? (The Problem of Inheritance of Acquired Characteristics)
DQ3. How can the male sexual element act not solely on the ovule, but occasionally on the mother-form? (The Problem of Sexual Influence)
DQ4. How a limb can be reproduced on the exact line of amputation, with neither too much nor too little added? (The Problem of Regeneration)
DQ5. How are the various modes of reproduction are connected? (The Problem of Reproduction)

Correspondingly, Darwin (1868, 374) also introduced the concept of gemmule and the hypotheses of pangenesis as the central constitutes of the solutions to his problems of heredity:

DT1. Cells propagate by self-division or proliferation to retain the same nature to become converted the tissues of the body.
DT2. Gemmules are the basic unit of hereditary material.
DT3. Gemmules are thrown off from cells, circulate freely throughout the organism, and become developed into cells again when supplied with proper nutriment during all the stages of development.
DT4. Gemmules are transmitted from generation to generation and generally developed in the immediate successive generation.
DT5. Some gemmules are in the dormant state in many generations and then developed.

Darwin (1868, 357) contended that these hypotheses were useful to solve his problems of heredity to explain both the sexual and asexual reproduction by "bringing together a multitude of facts which are at present left disconnected by any efficient cause." Although Darwin's hypotheses of pangenesis were heavily criticised by his contemporaries, his problems and solutions still marked a progress in the study of heredity by providing a useful, though imperfect or unsatisfactory, exemplary practice. In particular, Darwin's way of problem-defining and problem-solving was useful to classify the phenomena of inheritance and to complement his study of evolution. The explanation of atavism in terms of dormant germmules is a clear case.[1] Darwin's pangenesis approach was further developed by de Vries (1889). His main modifications, including the introduction of the distinction between active and latent pangens, the relocation of pangens, and the revision of the transportation of pangens, were conceptual. However, a persistent difficulty for both Darwin's and de Vries' theories was that there was no reliable way to test the hypothesis of pangenesis experimentally. (The problem of testability) Such a difficulty remained until 1900.

As I have shown in the previous chapters, at the turn of the twentieth century, Mendel's work on hybrid development (1866) was incorporated into a Mendelian theory, as alternative to the theory of pangenesis (Darwin 1868; de Vries 1889), in order to account for the mechanism of heredity. The major progress made by Mendel's theory is, as Bateson (1902, 2–3) pointed out, that it provided a reliable and testable way to study the "essential nature" of heredity. The theorists of heredity at the time, especially de Vries, were struggling to test their theories experimentally. Bateson insightfully recognised that Mendel's approach would be useful to solve this problem. As I have argued in Chap. 5, Mendel (1866, 4, 7) introduced a set of new research problems of transmission of morphological traits of peas by focusing on the paired traits in the successive generations.

MP1. How could one determine the number of different forms in which hybrid progeny appear, permit classification of these forms in each generation with certainty, and ascertain their numerical interrelationship?
MP2. What are the changes for each pair of differing traits in the offspring of *Pisum*? Or, what is the law deducible from the changes for each pair of differing traits selected in the successive generations?

Moreover, Mendel introduced various hybridisation experiments, the concepts of dominance and recessiveness, and the hypotheses (i.e. MH1, MH2, MH3) to solve the problems.

MH1. Of the seeds formed by the hybrids with one pair of differing traits, one half again develop the hybrid form while the other half yield plants that remain constant and receive the dominating and the recessive character in equal shares. (The law of development concerning a pair of differing traits)
MH2. The progeny of hybrids in which several essentially different traits are united represent the terms of a combination series in which the series for each pair of differing traits are combined. And the behaviour of each pair of differing traits in a hybrid association is independent of all other differences in the two parental plants. (The law of combination of differing traits)
MH3. Pea hybrids form germinal and pollen cells that in their composition correspond in equal numbers to all the constant forms resulting from the combination of traits united through fertilisation. (The law of composition of hybrid fertilising cells)

[1]The phenomenon of atavism was explained by DT5 in the sense that some gemmules were in the dormant state and remained undeveloped in many generations.

I have also argued in Chap. 5 that Mendel's research problems and solutions provided a reliable framework to study the mechanism of heredity experimentally. Thus, the progress marked by the shift from Darwin's approach to Mendel's approach can be well characterised by the increase of the usefulness of the research problems and the solutions. Mendel's exemplary practice was successfully repeated by Correns (1900) and Erich von Tschermak (1900a, 1900b). What is more, Mendel's problems of developmental series of the differing traits of peas and laws were used to solve the problem of testability, unsolved by Darwin and de Vries. While many of Darwin's hypotheses were difficult to test experimentally, Mendel's MH3 as a hypothesis of the mechanism of heredity was experimentally testable and confirmable. In addition, Mendel's laws and experiments were successfully applied to predict and explain the distribution of the differing morphological traits of various plant species[2] in the successive generations by de Vries (1900a, 87, 1900b, 846, 1903, 151) and Correns (1900, 160). Moreover, Mendel's exemplary practice generated more interdisciplinary and testable problems than the pangenesis approach. On the one hand, Mendel's problems and solutions provide a framework which is more experimentally testable than Darwin's. On the other hand, Mendel's exemplary practice was more interdisciplinary as it encompassed the studies of development and hybridisation. Thus, I argue that the progress made by Mendel's work is his more useful way of defining and solving the research problems.

Based on Mendel's approach, de Vries, Correns, and especially Bateson were developing the Mendelian approach to heredity by introducing new research problems, new concepts (e.g. anlage, unit-character), new hypotheses (e.g. de Vries' law of segregation, Correns' Mendelian rule, Bateson's Mendelian principles), and new experiments. The two main problems identified by the Mendelian approach were:

MQ1. What offspring are to be expected if two germ-cells of dissimilar constitution unite in fertilisation? (de Vries 1900b, 845; Bateson 1902, 18–19)

MQ2. Are the Mendelian principles universally applicable? (de Vries 1900b, 846–47; Correns 1900, 168; Bateson 1902, 33)

Compared with Mendel's approach, the Mendelian approach is more progressive in several aspects. Firstly, the concept of anlage (Correns 1900) and the concept of unit-character (Bateson 1902), by incorporating contemporary knowledge of cytology, provided a better understanding of the physical location of the bearer of hereditary material than the concept of cell-type (Mendel 1866). Cell-type was an inter-cellular element, while anlage and unit-character referred to some intra-cellular material within the nuclei, which were experimentally corroborated later. Secondly,

[2]The plant species displaying the Mendelian ratio in the hybridization experiment included *Agrostemma Githago, Amarantus caudatus, Aster Tripolium, Calliopsis tinetoria, Chelidonium majus, Chrysanthemum eoronarium, Clarhia pulchella, Corepis tinctoria, Datura Tabula, Hyosoyamus niger, Linaria vulgaris, Lychnis diurna, Lychnis vespertina, Oenothera Lamarckiana, Papaver somniferum Mephisto, Solanum nigrum, Trifolium pratense, Veronica longifolia, Viola cornuta,* and *Zea Mays.*

de Vries' law of segregation (1900b), Correns' Mendelian rule (1900), and Bateson's Mendelian principles (1902) were successfully applied to reinterpret and explain the hereditary pattern of more species, especially in human pedigrees (Gossage 1908; Nettleship 1909; Lundorg 1912, 1920). Thirdly, the Mendelian approach provided a framework to generate better defined research problems of heredity across the areas of heredity, hybridisation, animal breeding, cytology, evolution, and even medicine.

Therefore, now I can conclude that the progress in the study of heredity from Darwin to Bateson can be well characterised in terms of the increase of the usefulness of problem-defining and problem-solving.

6.4 The Problems of the Kuhn-Laudan Functional Approach Revisited

In this section, I argue that my new functional approach is better than the traditional functional approach. The most influential functional approach is first proposed by Thomas Kuhn (1962, 1970a), and mainly developed by Larry Laudan (1977, 1981).[3] This approach emphasises the significance of problem-solving. Kuhn (1970b, 164) argues that the nature of scientific progress is the increase of "both the effectiveness and the efficiency with the group as a whole solves new problems." Laudan (1981, 145) is also explicit on the point that "science progresses just in case successive theories solve more problems than their predecessors." Kuhn and Laudan differ in the explication of problem-solving. For Kuhn (1970b, 189–91), a problem *P* is solved if its solution is sufficiently similar to a relevant paradigmatic problem-solution. For Laudan (1977, 22–23), a problem *P* is solved by a theory *T* if *T* entails an approximate statement of *P*. Nevertheless, both Kuhn and Laudan maintain that scientific progress is nothing to do with truth or knowledge if truth or knowledge is construed in a classical way. More specifically, whether a problem is solved is independent of whether the paradigmatic solution assumes any paradigm-dependent truth (for Kuhn), or whether the background theory is true (for Laudan). As the acceptance of a problem solution is determined independently of external factors like truth or knowledge, whether a progress is achieved can be judged by the scientific community itself. Thus, the central tenets of the Kuhn-Laudan functional account of scientific progress can be summarised as follows.

KL1. Scientific progress is solely determined by the problem-solving power.
KL2. The problem-solving power is assessed by the amount and significance of the problems solved.

[3]Another representative of the functional approach is proposed and developed by Imre Lakatos. According to Lakatos (1978, 33–34), a research programme is progressive if it generates novel and well corroborated predictions. In this section, I focus on the Kuhn-Laudan functional approach, so I shall not delve into a detailed discussion on Lakatos' account.

KL3. The problem-solving power is independent of whether the solution is true or knowledge.
KL4. Scientific progress is judged and known by the scientific community.

There are two obvious problems of the Kuhn-Laudan functional approach. One is the problem of sufficiency. For Kuhn and Laudan, if a scientific community is working better and better on the effectiveness and efficiency of problem-solving, it implies scientific progress. On the contrary, if the effectiveness and efficiency of problem-solving decreases, then it marks a regress in science. The problem of sufficiency can be illustrated with a thought experiment proposed by Alexander Bird (2007, 69–70). Suppose there is a widely accepted but false theory. The scientific community accumulates the solutions to the problems derived from the false theory. This suggests progress, according to Laudan's approach. However, it seems implausible for many to accept that there is an ongoing progress in science, as the false solution statements (derived from the false theory) accumulate. What is worse, the Kuhn-Laudan approach is even more problematic when it comes to account for progress in a scientific revolution. Suppose at a later time t, the old false theory is replaced by a true theory. Bird argues that by Laudan's and Kuhn's standards,[4] the shift from the old false theory to a new true theory marks a regress, when the new theory solves fewer problems than the old one at the time t, as illustrated in Fig. 6.1. It is again implausible for many to accept that the shift from

Fig. 6.1 Progress in a scientific revolution

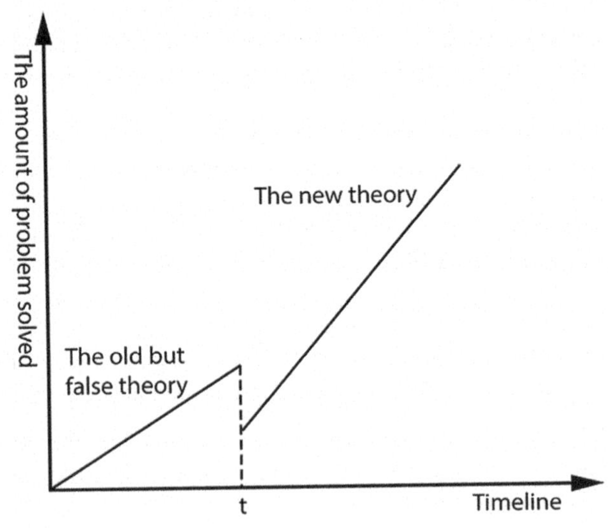

[4]Though Kuhn's criterion of puzzle-solving is distinct from Laudan's, Bird's thought experiment is applicable to Kuhn's approach by assuming that the paradigmatic solution relies on a false universal generalization.

a false theory to a true theory is a regress rather than a progress. Therefore, Bird concludes that the upshot of the Kuhn-Laudan approach to scientific progress is that radical scientific changes in the form of scientific revolutions are not progressive, and thus rejects KL1.

This conclusion is obviously not what Kuhn or Laudan would be happy to accept. Kuhn (1970b, 166) explicitly claims that "the outcome of revolution must be progress." For Kuhn, a new paradigm solves more problems than its predecessor. Thus, scientific revolutions are progressive. Nevertheless, Kuhn also well recognises the possibility that there is a loss of problem-solving power in some areas in a scientific change. Thus, it is an indispensable task for Kuhn and Laudan to work out a quantitative framework to justify that the increase of problem-solving power in some areas outweighs the loss in others. This is related to another obvious problem, namely, the problem of quantitative weighing, which is a persistent objection to the function approach (e.g. Collingwood 1965; Rescher 1984; Kleiner 1993): Is there a proper quantitative way to identify and calculate the problems of different significance? Laudan (1981, 149) proposes a straightforward formula to determine the problem-solving power: Scientific progress is achieved by maximising the number of important empirical problems solved and minimising the number of significant empirical anomalies and conceptual problems generated.[5] For example, a theory T_1 with 50 solved significant empirical problems and 10 significant empirical anomalies and conceptual problems has a better problem-solving power than T_2 with 10 solved significant empirical problems and 50 significant empirical anomalies and conceptual problems. Unfortunately, this strategy is still too oversimplified and vague to be helpful. Firstly, the significance of the empirical problems varies in degree. It is still unclear on how to evaluate the significance of the empirical problems in a quantitative way. Secondly, the significance of the anomalies is not obviously commensurate with that of the conceptual problems. Therefore, KL2 is seriously challenged by its feasibility.

In addition, there is another neglected problem, namely, the problem of internalism. According to Kuhn and Laudan, a scientific community well recognises whether it is making a progress or not by examining its problem-solving power. However, this is not the case in the history of science. The long neglect of Mendel's work in the study of heredity in the second half of the nineteenth century is such an example. Today Mendel's paper (1866) is widely accepted as the founding document of genetics. Nevertheless, its significance was not sufficiently recognised or appreciated until 1900, even if, as William Bateson (1902) pointed out, Mendel's paper did provide a reliable and promising framework to solve a persistent problem in the study of heredity in the nineteenth century: the "inward cause" (or the mechanism) of heredity. Therefore, it seems that the community of the study of

[5]Laudan distinguishes two types of scientific problems that are designed to be solved: empirical problems and conceptual problems. For Laudan (1977, 14–17), anything about the natural world in need of explanation is in the realm of empirical problems. Why heavy objects fall towards the earth is an empirical problem. In contrast, conceptual problems are all theory-dependent. What is absolute space is a conceptual problem.

heredity failed to recognise the progress made in Mendel's work, though it helped to solve key problems. But why? If Mendel's work was so progressive and important for the study of heredity, how could it be ignored or neglected for such a long time? Why was its significance not recognised earlier? And more generally, if scientific progress is judged and known by the community in terms of problem-solving power, how could those who studied heredity in the nineteenth century neglect the progressive feature of Mendel's work?[6] So, I argue that KL4 is shown to be problematic.

Moreover, the overlook of the functional approach in the recent literature arises from an intuition that the aim of science is about knowledge or truth. As the aim of science is widely assumed to be the acquisition of knowledge or the pursuit of truth, it seems natural to define scientific progress in terms of knowledge or truth. However, the functional approach usually eliminates the role of knowledge or truth in scientific progress. Laudan (1977, 126–27), for example, explicitly indicates that neither truth nor knowledge should ever be considered as the aim of science, and problem-solving is independent of truth or knowledge. Hence, the functional approach, especially KL3, is somehow counter-intuitive to many.

To sum up, all the four theses of the Kuhn-Laudan functional approach are under attack. Accordingly, there are four main problems of the Kuhn-Laudan functional approach: the problem of sufficiency, the problem of quantitative weighing, the problem of internalism, and the problem of counter-intuition.

I contend that my new functional approach well resolves the main problems of the Kuhn-Laudan functional approach. Firstly, I have argued that whether science progresses depends on whether a new way of defining and solving problems is more useful than the old one. In order to determine whether there is a progress in science, one has to examine whether a new way of problem-defining and problem-solving solves any problem unsolved by the old one, and whether that new way proposes more testable problems across more areas. Given such a qualitative notion of usefulness, there is no need to look for a quantitative framework to calculate and weigh the significance and amount of the problems. Thus, the problem of quantitative weighing is inapplicable to my functional approach. Secondly, my functional approach is not internalist. Bird construes both Kuhn's and Laudan's approaches as internalist in the sense that scientific progress is only judged and known by a community, independent of any features unknown to them. However, this does not apply to my approach. The usefulness of problem-defining and problem-solving is not straightforwardly recognisable by the scientific community, as I just illustrated in the Mendelian-Biometrician controversy. It is in this sense that my approach is not internalist. Therefore, KL4 is not assumed for my functional approach. Thirdly, I contend that by adding problem-defining as an essential constituent of scientific progress, the problem of sufficiency is resolved.[7]

Moreover, my functional approach differs from the Kuhn-Laudan functional approach in two main aspects. First of all, I highlight the significance of problem-defining. I agree with Kuhn (1970a, 1970b) and Laudan (1977) on the point that the

[6]For an exemplar-based explanation of the problem of the long neglect, see Chap. 7.

[7]The problem of counter-intuition shall be discussed in Sect. 6.5.

problems (or puzzles) are the focal point of scientific activities. Kuhn (1970b, 37) defines a puzzle as what is determined by the community to be solved, while Laudan (1977, 14) construes a problem as what is designed to be solved by theories. However, both Kuhn and Laudan implicitly assume that these problems are either simply pre-defined or defined in a straightforward way.[8] In other words, the significance of problem-defining seems not to be fully recognised. I argue that not all problems are simply defined in the ways that Kuhn and Laudan suggest. Problem-defining is much more than proposing a problem. In fact, it usually consists of activities of problem-proposing (i.e. to propose an initial problem), problem-refining (i.e. to refine an initial problem), and problem-specification (i.e. to make an initial problem into more specific and practical problems). Most, if not all, of the research problems are defined, refined, and redefined in the process of problem-solving. In contrast to Kuhn's and Laudan's view that science is essentially a problem-solving activity, I argue that science consists of both problem-defining and problem-solving activities.

Secondly, my concept of problem-solving is different from Kuhn's puzzle-solving or Laudan's problem-solving. As I mentioned, Kuhn suggests that puzzle-solving is an activity of looking for a solution which is sufficiently similar to a relevant paradigmatic problem-solution, while Laudan argues that a problem P is solved by a theory T if T entails an approximate statement of P. In contrast, I do not think that the constituents of the solution to a research problem can be characterised in a monistic way. Scientists solve the problems in different ways, so it would be unwise for anyone to try to summarise some universally fundamental characteristics of their solutions. As I have argued in Chap. 5, problem-solving is a series of intertwined activities of experimentation, problem-refining, conceptualisation, hypothesisation, and reasoning. It is evident that these components are intertwined. For instance, the hypotheses are often formulated on the basis of the results of the experiments by employing the concepts in the vocabulary; the experiments are usually designed and undertaken with the purpose of solving the research problems (e.g. by testing the hypotheses); the concepts in the vocabulary are understood with the help of undertaking the experiments and applying the hypotheses, and so on.

6.5 Beyond Knowledge, Truth, and Intervening

There is still one more problem, namely, the problem of counter-intuition, yet to be discussed. As I mentioned in Sect. 6.4, the functional approach is somehow neglected in the recent debate for its conflict with the intuition that scientific progress is about knowledge or truth. However, I argue that my functional approach can be

[8]For Kuhn (1970b), puzzles are defined (or even pre-defined) relative to disciplinary matrices which assure the existence of their solutions. For Laudan (1977), empirical problems consist of unsolved problems, solved problems, and anomalous problems, He says little on how unsolved problems and solved problems are defined, and claims that anomalous problems often are generated by new observations.

compatible with this intuition. Firstly, the usefulness of problem-defining and problem-solving could be well explained in terms of knowledge if knowledge is not merely construed as something propositional or theoretical. Knowledge is traditionally classified into know-that and know-how.[9] The way of defining and solving the problems can be understood as a case of know-how. Thus, that more useful problems and their solutions are proposed could be understood in the sense that more useful know-how is obtained.

Secondly, the usefulness of problem-defining and problem-solving is also explicable in terms of truth to some extent. In particular, it is well explained by the "contextualist" theory of truth (Chang 2012; Massimi 2018). Michela Massimi (2018), for example, proposes that truth in the context of scientific practice should be defined in a perspectival way.

> Knowledge claims in science are [perspective-dependent] when their *truth-conditions* (understood as rules for determining truth-values based on features of the context of use) depend on the scientific perspective in which such claims are made. Yet such knowledge claims must also be assessable from the point of view of other (subsequent or rival) scientific perspectives. (Massimi 2018, 354)

If truth is defined in this perspectival way, then the increase of the usefulness of problem-defining and problem-solving implies the increase of unsolved problems with more useful hypotheses as the solutions. The Mendelian approach to the study of heredity, for example, generates more confirmable hypotheses (e.g. the law of segregation) and factual knowledge (e.g. the summary of the transmission of morphological traits of various plants). All these hypotheses and factual knowledge are shown to be true according to the Mendelian perspective by means of experiments, while they are also assessable from the point of view of the subsequent scientific perspective (e.g. the Morgan approach) by new ways of experimentation. Therefore, the concept of more useful problems and their solutions being proposed could be interpreted as more perspective-dependent true knowledge claims are attained.[10]

Furthermore, I argue that my functional account is better than the epistemic and semantic accounts in one significant aspect. The epistemic and semantic approaches to scientific progress pay too much attention to theoretical achievements. However, such approaches overlook the significance of the non-theoretical aspect of science. As Heather Douglas (2014, 56) points out, science is not just about theory, and thus scientific progress should be examined in both theoretical and non-theoretical aspects. One advantage of my functional account is to highlight the significance of the non-theoretical aspect of scientific progress. Scientific practice is much more

---

[9]Jason Stanley and Timothy Williamson (2001) famously rejects this distinction by arguing that know-how is reducible to know-that. Whether there is a genuine distinction between know-that and know-how, my point still holds. Science does not only tell us something theoretical which can be formulated in the propositions, but also tell us something practical, whether which can be reformulated in the propositions or not.

[10]It should be highlighted that the notion of usefulness can be explicated by the contextualist theory of truth does not imply that my functional approach assumes a contextualist theory of truth. It does not eliminate the possibility that it can also be explicated by other theories of truth.

than theorising or modelling. It would be surprising if the introduction of new research problems and the improvement of the experimental methods and devices are excluded from the constituents of scientific progress. Reconsider the case of the origin of genetics. It seems plausible to argue that the progress made by Mendel was the proposal of the law of composition of hybrid fertilising cells (MH3) which advanced our knowledge of the mechanism of heredity. Similarly, de Vries' law of segregation, Correns' Mendelian rule, and Bateson's Mendelian principles provide a better knowledge of heredity than Mendel's law. Yes, we knew more and more about the mechanism of heredity with the theoretical development from Darwin to Bateson. However, it is definitely not the only aspect of the progress achieved in the study of heredity in that period. We learnt more and more on how to define a good research problem, how to design and undertake the experiments, and how to use the problems and experiments to study the mechanism of heredity in a better way. All of these cannot easily be accounted for in terms of theoretical knowledge by the epistemic approach or in terms of truth by the semantic approach. The non-theoretical aspect of scientific progress should be taken into account as well as the theoretical aspect. Thus, I contend that my functional approach provides a better account for the non-theoretical aspect of scientific progress.

Now there are two potential objections to my argument that my functional approach is better than the epistemic and semantic ones. Firstly, as my reply to the problem of counter-intuition involves the notions of truth and knowledge, one may wonder whether my approach can be still classified as "functional" rather than epistemic or semantic. It seems that my functional approach can be reinterpreted in the way that science progresses if more and more know-how is attained, or that science progresses if more perspective-dependent truths are obtained. Secondly, one may argue that the overlook of the non-theoretical aspect of scientific progress is not a serious problem for the epistemic or the semantic approach. There is lots of know-how in conceptualisation and experimentation, but this kind of progress is typically treated as secondary. The theoretical aspect of scientific progress is more fundamental as it explains the non-theoretical aspect.

In response to the first potential objection, I argue that my functional approach is not reducible to the epistemic or the semantic approach. As I have pointed out, the solution to a research problem is not something purely theoretical or propositional. There are some indispensable non-theoretical aspects. Mendel's solution to MP1 and MP2, for example, consists of a series of practical guides on how to design and undertake the experiments. Accordingly, I do not see that there is any true or correct solution to a research problem, given its practical nature. For example, it is implausible to claim that there is a true or correct way of experimentation or problem-refining. Moreover, I would like to highlight that while my approach aims to capture the different aspects of scientific progress, it does not imply that the functional, epistemic, and semantic aspects constitute the nature of scientific progress in equal shares. Rather more know-how (the epistemic aspect) and more well-corroborated hypotheses (the semantic aspect) only partially constitute the usefulness of problem-defining and problem-solving (the functional aspect). It is in this sense that my approach is functional rather than epistemic, semantic, or integrated.

Regarding the second potential objection, I have to emphasise that the theoretical aspect of the progress is in no sense more fundamental than the non-theoretical one. Hypothesisation is mutually intertwined with and cannot be independent of other activities of problem-defining and problem-solving. The theoretical aspect of scientific progress is achieved with the non-theoretical aspect. Mendel's work cannot mark any progress without his novel problem-defining, conceptualisation, and experimentation, which cannot be completely accounted for in terms of knowledge or truth. Therefore, I argue that, as the non-theoretical aspect of scientific progress is indispensable and irreducible, the neglect of the non-theoretical aspect of scientific progress is a serious problem for the epistemic and the semantic approaches. Neither the epistemic nor the semantic approach provides a complete account of scientific progress. In addition, even if the usefulness of problem-defining and problem-solving can be understood in terms of knowledge or truth to some extent, the mere accumulation of scientific knowledge or the approximation of scientific truths, which do not contribute to the increase of the usefulness of problem-defining and problem-solving, do not count as scientific progress. I argue that the accumulation of scientific knowledge or the approximation of scientific truth is usually a result rather than an indicator of scientific progress. Therefore, it is in this sense that my approach is fundamentally functional rather than epistemic or semantic.

It should be noted that Douglas (2014, 62) proposes a functional account of scientific progress in terms of "the increased capacity to predict, control, manipulate, and intervene in various contexts." There are some similarities between Douglas and my approach. Both highlight the non-theoretical aspect of scientific success. Both admit that scientific progress is not a concept which can be easily quantified. At first glance, the origin of genetics from Darwin to Bateson could also be explained and characterised in terms of increasing capacity to predict, control, manipulate, and intervene. Mendel's approach had a more increased capacity to predict, control, manipulate, and intervene the transmission of the morphological traits of *Pisum* than Darwin's, and the Mendelian approach had an even more increased capacity to predict, control, manipulate, and intervene in the transmission of the morphological traits of various species than Mendel's. However, I have to highlight that there are some substantial differences between Douglas' and my approach (2014, 62). Compared with my approach, Douglas' approach overlooks the significance of the theoretical aspect of scientific progress to some extent. Not all the theoretical progress is adequately reflected in the increase of the capacity to predict, control, manipulate, and intervene. De Vries, for example, spent two decades developing Darwin's theory of pangenesis. Although it is still controversial if de Vries' approach had any increased capacity to predict, control, manipulate, and intervene in the transportation of pangens than Darwin's, it is undeniable that de Vries made substantial conceptual revisions of the pangenesis theory to accommodate the development of cytology at the time. It is implausible to argue that de Vries' approach to the problem of heredity was not progressive, especially given that it did advance the understanding of the physical location of hereditary material by identifying that the pangens are located in the nuclei of cells. In addition, Douglas' approach pays insufficient attention to the significance of the introduction of new research problems, concepts, hypotheses, and experiments in practice. It seems plausible to characterise the progress made by Mendel's approach as the increased capacity to predict, control, manipulate, and intervene the transmission of the morphological traits of *Pisum*, but such an increased capacity is achieved by Mendel's way of defining new research problems, introducing new concepts (e.g. dominance), proposing new hypotheses (e.g. MH1, MH2, and MH3), and designing and undertaking new experiments. Douglas might argue that Mendel's problem-defining and problem-solving consist in the capacity for this. Nevertheless, I argue that Mendel's problem-defining and problem-solving are more fundamental in accounting for the nature of scientific progress. Mendel's problem-defining and problem-solving underlie the capacity to predict, control, manipulate, and intervene in the transmission of the morphological traits of *Pisum*. It is Mendel's problem-defining and problem-solving rather than Mendel's capacity to predict, control, manipulate, and intervene in the transmission of the morphological traits of *Pisum* that guided de Vries', Correns', and Bateson's work on heredity. Thus, the increase of the usefulness of Mendel's problem-defining and problem-solving implies the increase of the capacity to predict, control, manipulate, and intervene in the transmission of the morphological traits of *Pisum* rather than that the increased capacity to predict, control, manipulate, and intervene in the transmission of the morphological traits of *Pisum* implies the increase of the usefulness of Mendel's problem-defining and problem-solving.

Before finishing this section, I would like to ward off a common misunderstanding of the functional approach to scientific progress. Bird (2007), for example, regards accepting a functional approach as rejecting scientific realism or endorsing antirealism. However, I argue that a functionalist on scientific progress does not have to an anti-realist. My functional approach is neutral to the scientific realism/antirealism debate. Bird argues that the acceptance of the pessimistic metainduction argument and the rejection of transcendental truth are two sources of the functional approach. Nevertheless, the acceptance of the pessimistic metainduction argument and the rejection of transcendental truth do not necessarily imply an anti-realist position. John Worrall's structural realism (1989) is a good such example. Worrall is explicit on the point that he accepts the historicist challenge raised by the pessimistic metainduction argument, but it does not undermine his realist position. Both realists and antirealists can take the functional approach to scientific progress, though they may differ in making the ontological commitment to the hypotheses. In addition, I do not think that my functional approach to scientific progress is dependent on the acceptance of the pessimistic metainduction argument and the rejection of transcendental truth. What the pessimistic metainduction rejects is a cumulative and convergent picture of the development of science. None of the epistemic, the semantic, or the functional approach implies a particular historiography of science. What my functional approach, as well other approaches, provides is an account of the nature of scientific progress. Similarly, how to define truth is not an indispensable task for my functional approach. In other words, my functional approach is neutral to the definition of truth. Nor does it endorse or reject any particular theory of truth.

6.6 Conclusion

In this chapter, motivated by the exemplar-based approach, I have developed a new functional approach to scientific progress. I have argued that the progress in early genetics can be well characterised in terms of the increase of the usefulness of problem-defining and problem-solving. Moreover, I have argued that the new functional approach is better than other approaches to scientific progress generally.

References

Bateson, William. 1902. *Mendel's Principles of Heredity: A Defence*. Cambridge: Cambridge University Press.

Bird, Alexander. 2007. What is Scientific Progress? *Noûs* 41 (1): 64–89.

Chang, Hasok. 2012. *Is Water H_2O? Evidence, Realism and Pluralism*. Dordrecht: Springer.

Collingwood, Robin George. 1965. *The Idea of History*. Oxford: Oxford University Press.

Correns, Carl. 1900. G. Mendels Regel über das Verhalten der Nachkommenschaft der Rassenbastarde. *Berichte der Deutschen Botanischen Gesellschaft* 18 (4): 158–168.

Darwin, Charles. 1859. *On the Origin of Species*. London: John Murray.

———. 1868. *The Variation of Animals and Plants under Domestication*. London: John Murray.

Douglas, Heather. 2014. Pure Science and the Problem of Progress. *Studies in History and Philosophy of Science* 46: 55–63.

Galton, Francis. 1889. *Natural Inheritance*. London and New York: Macmillan & Company.

Gossage, A.M. 1908. The Inheritance of Certain Human Abnormalities. *Quarterly Journal of Medicine* 3 (1): 331–347.

Hart, D. Berry. 1909. Mendelian Action on Differentiated Sex. *Transactions of the Edinburgh Obstetrical Society* 34: 303–357.

Holmes, S.J., and H.M. Loomis. 1909. The Heredity of Eye Color and Hair Color in Man. *Biological Bulletin* 18 (1): 50–56.

Kleiner, Scott A. 1993. *The Logic of Discovery: A Theory of the Rationality of Scientific Research*. Dordrecht: Kluwer Academic Publishers.

Kuhn, Thomas Samuel. 1962. *The Structure of Scientific Revolutions*. 1st ed. Chicago, IL: The University of Chicago Press.

———. 1970a. Logic of Discovery or Psychology of Research? In *Criticism and the Growth of Knowledge*, edited by Imre Lakatos and Alan Musgrave, 1–23. Cambridge: Cambridge University Press.

———. 1970b. *The Structure of Scientific Revolutions*. 2nd ed. Chicago, IL: The University of Chicago Press.

Lakatos, Imre. 1978. Falsification and the Methodology of Scientific Research Programmes. In *The Methodology of Scientific Research Programme*, edited by John Worrall and Gregory Currie, 8–101. Cambridge: Cambridge University Press.

Laudan, Larry. 1977. *Progress and Its Problems: Toward a Theory of Scientific Growth*. Berkeley and Los Angeles: The University of California Press.

———. 1981. A Problem-Solving Approach to Scientific Progress. In *Scientific Revolutions*, edited by Ian Hacking, 144–155. Oxford: Oxford University Press.

Lundorg, Herman Bernhard. 1912. Ueber die Erblichkeitsverhältnisse der Konstitutionellen (Hereditären) Taubstummheit und Einige Worte über die Bedeutung der Erhlichkeitsforschung für die Krankheitslehre. *Archiv für Rassen- und Gesellschafts-Biologie* 9.

———. 1920. Hereditary Transmission of Genotypical Deaf-Mutism. *Hereditas* 1 (1): 35–40.

Massimi, Michela. 2018. Four Kinds of Perspectival Truth. *Philosophy and Phenomenological Research* 96 (2): 342–359.
McCracken, Isabel. 1905. A Study of the Inheritance of Dichromatism in Lina Lapponica. *Journal of Experimental Zoology* 2 (1): 117–136.
———. 1906. Inheritance of Dichromatism in Lina and Gastroidea. *Journal of Experimental Zoology* 3 (2): 321–336.
———. 1907. Occurrence of a Sport in Melasoma (Lina) Scripta and Its Behavior in Heredity. *Journal of Experimental Zoology* 4 (2): 221–238.
Mendel, Gregor. 1866. Versuche über Pflanzenhybriden. *Verhandlungen des Naturforschenden Vereins Brünn* IV (1865) (Abhandlungen): 3–47.
Nersessian, Nancy J. 2008. *Creating Scientific Concepts*. Cambridge, MA: The MIT Press.
Nettleship, Edward. 1909. *The Bowman Lecture on Some Hereditary Diseases of the Eye*. London: Adlard and Son.
Niiniluoto, Ilkka. 2014. Scientific Progress as Increasing Verisimilitude. *Studies in History and Philosophy of Science* 46: 73–77.
Prout, Louis B. 1907. "Xanthorhoe Ferrugata (Clark) and the Mendelian Hypothesis." *Transactions of the Entomological Society of London for the Year 1906*, 525–31.
Reid, George Archdall. 1905. *The Principles of Heredity with Some Applications*. New York: E. P. Dutton & Co.
Rescher, Nicholas. 1984. *The Limit of Science*. Berkeley, CA: The University of California Press.
Saunders, Charles E. 1907. The Inheritance of Awns in Wheat. In *Report of the Third International Congress on Genetics 1906*, 370–372. London: The Royal Horticultural Society.
Stanley, Jason, and Timothy Williamson. 2001. Knowing How. *The Journal of Philosophy* 98 (8): 411–444.
von Tschermak, Erich. 1900a. Über Künstliche Kreuzung bei Pisum Sativum. *Berichte der Deutschen Botanischen Gesellschaft* 18 (6): 232–239.
———. 1900b. Über Künstliche Kreuzung bei Pisum Sativum. *Zeitschrift für das Landwirtschaftliche Versuchswesen in Oesterreich* 3: 465–555.
de Vries, Hugo. 1889. *Intracellulare Pangenesis*. Jean: Gustav Fischer.
———. 1900a. Das Spaltungsgesetz der Bastarde (Vorlaufige Mittheilung). *Berichte der Deutschen Botanischen Gesellschaft* 18 (3): 83–90.
———. 1900b. Sur la Loi de Disjonction des Hybrides. *Comptes Rendus de l'Academie des Sciences (Paris)* 130: 845–847.
———. 1903. *Die Mutationstheorie (II)*. Lepzig: Verlag von Veit & Comp.
Weldon, Walter Frank Rapheal. 1902a. Mendel's Laws of Alternative Inheritance in Peas. *Biometrika* 1 (2): 228–254.
———. 1902b. On the Ambiguity of Mendel's Categories. *Biometrika* 2 (1): 44–55.
Whitman, C. O. 1904. Hybrids from Wild Species of Pigeons, Crossed Inter Se and with Domestic Races. *Biological Bulletin* 6 (6): 315–316.
Worrall, John. 1989. Structural Realism: The Best of Both Worlds? *Dialectica* 43 (1–2): 99–124.

Chapter 7
The Problem of the Long Neglect Revisited: An Exemplar-Based Explanation

Abstract It is a puzzle why the significance of Mendel's paper *Versuche über Pflanzen-Hybriden* for the study of heredity was not well appreciated until the 1900s. This chapter revisits the problem of the long neglect and provides a new philosophical explanation. I begin with a distinction between two versions of the problem. I critically examine the traditional diagnoses of the problem. Then I defend the exemplar-based account of Mendel's contribution. Finally, I further develop an exemplar-based explanation of problem of the long neglect.

Keywords Mendel · Exemplary practice · Exemplar-based approach · Usefulness · Repeatability

7.1 Two Problems of the Long Neglect

In the last two chapters, I have argued that the origin of genetics from Gregor Mendel to William Bateson should be best characterised as a chain of exemplary practices, and the progress in early genetics can be well characterised as the increase of the usefulness of problem-defining and problem-solving. In this chapter, I argue that the exemplar-based analysis of the origin of genetics sheds new light on a persistent puzzle in the history and philosophy of biology, the problem of the long neglect.

The puzzle was rooted in a once widespread historiography: Mendel's paper (1866) was neglected by his contemporaries until 1900 when Hugo de Vries, Carl Correns, and Erich von Tschermak rediscovered it. However, such a historiography is highly problematic. Mendel's paper was not completely forgotten or ignored. Some studies (e.g. Olby 1985, 219–34; Weiling 1991, 10–11; Orel 1996, 275–79) have shown that there were at least a dozen references to Mendel's paper before 1900. In particular, Mendel's paper was carefully studied and discussed by C. A. Blomberg (1872) and I. F. Schmalhausen (1874) in their theses. So, it is clearly not the case that Mendel's paper was completely neglected or unknown in academia. If Mendel's paper was not neglected, then the nature of the long neglect seems to be unclear. What was neglected if there is anything? There are various phrases of the

Y. Shan, *Doing Integrated History and Philosophy of Science: A Case Study of the Origin of Genetics*, Boston Studies in the Philosophy and History of Science 320,
https://doi.org/10.1007/978-3-030-50617-9_7

rediscovery story: "the rediscovery of Mendel's work" (e.g. Punnett 1919; Iltis 1932; Corcos and Monaghan 1985; Orel 1996), "the rediscovery of Mendelism" (Stern and Sherwood 1966; Zirkle 1968; Harwood 2000), and "the rediscovery of Mendel's law (s)" (Bateson 1902; Stubbe 1972; Kottler 1979; Bowler 1989; Simuneket al. 2011). In fact, these phrases are sometimes used interchangeably (Tschermak 1951; Corcos and Monaghan 1985, 1987a, 1987b; Monaghan and Corcos 1986, 1987). Conventionally, both Mendel's work and Mendelism are identical with Mendel's law of heredity. So, it suggests that what was neglected was Mendel's law of heredity. Clearly, this contradicts our best historiography of Mendel. As I have shown in Chap. 2, it is now a consensus that Mendel's work on *Pisum* was about development (*Entwicklung*) rather than heredity (*Vererbung*). In addition, Mendel did propose three laws, but none of them was named law of heredity. In other words, there was no such a thing as Mendel's law of heredity in Mendel's paper. Thus, it is inappropriate to construe the nature of the long neglect as the neglect of Mendel's law of heredity. Then, what about Mendel's laws of hybrid development? It is again dubious. De Vries did not mention Mendel's laws in his rediscovery papers (1900a, 1900b, 1900c) at all, though he gave credit to Mendel for the discovery of the law of segregation (de Vries 1900a, 85). Correns (1900) formulated Mendel's rule,[1] but it was substantially different from Mendel's laws. Tschermak never explicitly stated what Mendel's principle was. If Mendel's laws were what were neglected, then it is not obvious that their significance was appreciated in 1900 in a strict sense. Nevertheless, most would agree that Mendel's work was overlooked to a great extent by those who studied the problem of heredity in the late nineteenth century, and its significance for the study of heredity was not widely appreciated until the first decade of the twentieth century.

Therefore, it should be highlighted that there are two versions of the problem of the long neglect of Mendel's work. One version is: Why was Mendel's work neglected in the late nineteenth century at all? (The problem of the long neglect of Mendel's work) The other version is: *Why was the significance of Mendel's work for the study of heredity overlooked in the late nineteenth century*? (The problem of the long neglect of the significance of Mendel's work for the study of heredity) The former version of the problem is largely a pseudo-problem. The long neglect in this sense is, as L. A. Callender (1988, 72) points out, "a product of historians of science, not of scientific history." However, the latter version of the problem is a real puzzle for historians and philosophers of science. It is undeniable that Mendel's work, though not about heredity, made a great impact on the early development of genetics. But why its significance for the study of heredity was suddenly recognised at the turn of the twentieth century? Why not earlier? Thus, the central question of this chapter

[1]The title of Corren's paper (i.e. "G. Mendels Regel über das Verhalten der Nachkommenschaft der Rassenbastarde") is usually translated as "G. Mendel's Law concerning the Behavior of Progeny of Varietal Hybrids." In particular, *Mendel'sche Regel* is translated by Piternick (1966) as "Mendel's law." However, this translation is misleading, as *Regel* is better translated as "rule" rather than "law."

is: Why was the significance of Mendel's work for the study of heredity not recognised earlier?

In this chapter, I argue that the exemplar-based account of the origin of genetics provides a new philosophical (or intellectual) explanation of the long neglect of the significance of Mendel's work for the study of heredity. In Sect. 7.2, I critically examine the traditional diagnoses of the problem. In Sect. 7.3, I defend the exemplar-based account of Mendel's contribution. In Sect. 7.4, I further develop an exemplar-based explanation of problem of the long neglect.

7.2 The Traditional Diagnoses of the Long Neglect

There are two popular explanations of the long neglect: Mendel's work was not accepted by those who were studying the problem of heredity in the late nineteenth century; Mendel's work was not well known by his nineteenth century contemporaries. Surely, these explanations are not fully satisfactory. We also need further explanations: Why was Mendel's work not accepted? Why was Mendel's work unknown? In the following sections, I shall examine the traditional explanations of these issues.

7.2.1 *Explanation 1: Mendel's Work Was Not Accepted*

Hugo Iltis (1932) famously argues that during the mid-nineteenth century the time was not ripening for understanding Mendel's work. But finally, at the turn of the twentieth century, "his time has come". (*The Problem of Prematurity*)

> [T]he study of the cell nucleus during the closing decades of the nineteenth century, thanks to which the chromosomes had come to be recognised as the bearers of heredity, while the reduction division had been observed and its purpose understood, and fertilisation (amphimixis) had been recognised as one of the most important causes of variability; Johannsen's experiments, and this investigator's vivid formulation of "population" and "pure line," of "phenotype" and "genotype"; de Vries' discovery of mutation, and the conception he based upon the mutation theory that a species is a mosaic of characters. Thanks to these converging trends, by 1900 the scientific situation was such that experiments of the kind performed thirty-five years earlier by Mendel had become absolutely essential to the testing of the various theoretical views. (Iltis 1932, 301–2)

In short, for Iltis, Mendel's work was ahead of his time. The significance of Mendel's work was not recognisable until the further development in the fields like cytology and heredity. This explanation had been widely accepted. There are some similar views.

> Mendel had produced a key piece for the jigsaw of biological theory – a much more important piece than he could have realized – but it was of no general use until the picture was sufficiently complete for it to be fitted in. (Gasking 1959, 77)

> One may say that the criterion of prematurity,[2] as defined by Stent, without question applies well to the class cases of neglect of the work of Mendel . . . (Glass 1974, 110)

In addition, the lack of generality of Mendel's laws is another explanation. (*The Problem of Lack of Generality*) It is clear that even Mendel (1866, 42) himself well realised that his laws were not universally applicable. Thus, some historians regarded this as a key factor of the lack of success of Mendel's work.

> Although Mendel's theory in fact ordered a mass of existing empirical knowledge, its application to this was not immediately and easily apparent. The theory must have seemed to have no other evidence in its favour than that collected by Mendel himself. (Wilkie 1962, 5)

> The analysis also suggests strongly that another criterion should be added to the determinative process, that of a lack of generality. Mendel's peas were thought to be unsuitable material for studying the heredity of species difference, and his laws were not clearly applicable to the other plants he attempted to use. (Glass 1974, 110)

Mendel's mathematical approach is also regarded as an obstacle for his contemporaries to understand and appreciate his work. (*The Problem of Obscurity of Mendel's Mathematical Approach*)

> [Mendel] was really a physicist, and brought to one of the great problems of biology the attitude of mind and the quantitative method of attack which had been in use for some time by physicists and by astronomers, and which was just coming to be used more widely by chemists. It was an unknown language to biology, though it fulfilled the essential requirements of scientific research better than anything which had gone before; and it came to biology at a time when those who were endeavouring to investigate inheritance by means of hybridization were not prepared for their task. (East 1923, 232)

> Mendel's mathematical treatment of his botanical data must have seemed strange in that time, when quantitative biology was unheard of. (Dodson 1955, 194)

> [B]iology was not ready for mathematical treatment. (Weinstein 1962, 999)

Moreover, Mendel's concern seems to have been removed from the interests of both biologists and breeders at his time. (*The Problem of Irrelevance*) In other words, Mendel's work was thought to be irrelevant to the central problems in the study of heredity in the nineteenth century.

> Mendel's contemporaries therefore tended, either to misinterpret his work as a confused attempt to investigate the nature of species, or else to dismiss it as being irrelevant to their own crucial problem of the origin of species. (Gasking 1959, 61)

> [A] more general reason for the neglect of Mendel's work . . . that Mendel's contemporaries were unable to see the importance of his ideas because these ideas did not seem to them have any relevance to the problem of the nature and origin of species, which at that time appeared to be the central problem of biology. (Wilkie 1962, 5)

---

[2]Gunther Stent (1972, 84) provides a criterion of the prematurity of a scientific discovery to explain its failure to make an impact in its time: "A discovery is premature if its implications cannot be connected by a series of simple logical steps to canonical, or generally accepted, knowledge."

Laurence Sandler and Iris Sandler (1985) also provide an explanation of why "Mendel's contemporaries failed to understand Mendel's work properly": Mendel's usage of the term *Entwicklung* was so novel that it was hardly useful to be applied in the prevailing conceptual framework in the mid nineteenth century. (*The Problem of Novelty*) At the time, there was no well-defined problem of transmission of morphological traits at all. The distinction between the problem of heredity and the problem of development was fuzzy. Most of Mendel's contemporaries, for example, Charles Darwin, accepted a developmental model of heredity in which the transmission of morphological traits from one generation to the next and the process by which the morphological traits are produced in the growing organisms are just different stages of the same biological process. According to Sandler and Sandler, it was Mendel who first explicitly defined the problem of transmission, the significance of which was not ever recognised. Peter Bowler (1989, 108–9) further argues, "For [Mendel's] paper to become useful, however, it had to be read in a context very different to that in which it had been written . . . Only when that framework changed would it become possible for a new generation of biologists to look back and see that the inheritance of characters in peas offered a model upon which to build a whole new theory of heredity." Thus, Mendel's work was difficult to be accepted by his contemporaries, and its significance for the study of heredity failed to be appreciated until 1900.

7.2.2 Explanation 2: Mendel's Work Was Unknown

It has been argued that Mendel was an unknown monk and published a paper in an obscure journal at the time. Moreover, Mendel did not have any collaborators or students to follow his work. Even worse, some (e.g. Mayr 1982) have pointed out that Mendel's characteristic modesty prevented him from advertising his work to others. All these factors lead to the inaccessibility of Mendel's paper.

In addition, Bateson suggests that a key factor of the inaccessibility of Mendel's work is that it was part of the works of hybridism, which was not the mainstream research school in the study of heredity.

> It is true that the journal in which [Mendel's paper] appeared is scarce, but his circumstance has seldom long delayed general recognition. The cause is unquestionably to be found in that neglect of the experimental study of the problem of Species which supervened on the general acceptance of the Darwinian doctrines. The problem of Species, as Kölreuter, Gärtner, Naudin, Wichura, and the other hybridists of the middle of the nineteenth century conceived it, attracted thenceforth no workers. (Bateson 1902, 37)

Michael MacRoberts (1985) suggests another factor of the inaccessibility of Mendel's work. By comparing the fate of Mendel's paper on *Pisum* with the one on *Hieracium* on the one hand, and the informal communication network around these two papers on the other hand, he finds that the reception of Mendel's papers well confirms William Garvey and Belver Griffith's psychological study (1971) that most scientific communication takes place at the informal level. Therefore,

MacRoberts argues that one significant explanation of the long neglect problem is that Mendel made too little informal communication with his contemporaries.

7.2.3 Summary and Remarks

These traditional explanations are all plausible to some extent. For example, there was a problem of lack of generality of Mendel's work on hybrid development, especially his laws (MH1, MH2). As Carl Correns (1900, 168, 1966, 132) indicated, "Mendel's [rule][3] of segregation cannot be applied universally."[4] In addition, because of his novel approach to developmental series,[5] Mendel's work was not easy to be comprehended by his contemporaries. Mendel's work, though not completely ignored, was not well known at the time (Table 7.1).

But these traditional explanations still fail to provide a complete account of the problem of the long neglect in the sense that most of them only provide the accounts of why the significance of Mendel's work was neglected. However, I argue that a complete explanation of the long neglect should encompass not only an account of why the significance of Mendel's work was neglected in the late nineteenth century, but also an account of why the significance of Mendel's work was appreciated at the

Table 7.1 The traditional explanations of the long neglect

Explanation	Further explanation
Mendel's work was not accepted by his contemporaries.	Ahead of his time / Prematurity (Iltis 1932; Gasking 1959; Zirkle 1964; Glass 1974)
	Lack of generality (Wilkie 1962; Glass 1974)
	Obscure application of mathematical models (East 1923; Dodson 1955; Gasking 1959; Weinstein 1962; Wilkie 1962)
	Irrelevance to the central problems of his time (Wilkie 1962; Gasking 1959)
	Failure of comprehension (Gasking 1959; Sorsby 1965; Sandler and Sandler 1985)
Mendel's work was to a large extent unknown by his contemporaries.	Lack of informal communication (MacRoberts 1985)
	Distraction by Darwin's work (Bateson 1902)
	Inaccessibility of Mendel's paper (Dorsey 1944; Dodson 1955; Posner and Skutil 1968; Mayr 1982)

[3]For the reason stated in footnote 1 of this chapter, I contend that *Mendel'sche Spaltungsregel* should be better translated as "Mendel's rule of segregation" rather than "Mendel's law of segregation".

[4]Mendel's rule here refers to CH2, a reformulated version of MH3.

[5]For a detailed discussion on Mendel's "developmental series", see Sect. 2.3. And see Sect. 5.5 for an exemplar-based analysis of Mendel's work.

turn of the twentieth century. There was a problem of lack of generality of Mendel's work in the nineteenth century, and the problem remained at the turn of the twentieth century. Why was the lack of generality an obstacle for Mendel's work to be neglected in the late nineteenth century but no longer a problem at the turn of the twentieth century? In other words, if the problems, which contributed to the neglect of the significance of Mendel's work, remained around 1900, why was its significance suddenly recognised? Therefore, I argue that a new examination of the problem of the long neglect is necessary in order to solve this puzzle.

Two questions are central to the new examination: What was the significance of Mendel's work for the study of heredity? Why was it neglected? There is another way to formulate the first question: What was Mendel's contribution to the study of heredity? Accordingly, the second question will be reformulated as why Mendel's contribution was neglected. Thus, my strategy is to begin with an analysis of Mendel contribution. Then I shall examine the reasons why Mendel's contribution was overlooked.

7.3 Mendel's Contribution Reconsidered

Today, historians and Philosophers have different approaches to Mendel's contribution, as they have different perspectives and interests. Although I aim at a philosophical (or intellectual) account, I contend that a good philosophical (or intellectual) account should be well collaborated by our best historiography.

7.3.1 The Traditional Philosophical Analyses

The most popular account of Mendel's contribution, still found in the genetics textbooks, is that Mendel introduced a theory of heredity (Suzuki et al. 1981; Brown 1989; Hartl et al. 1989; Atherly et al. 1999; Hartwell et al. 2008). There are two obvious mistakes in this account. As I have argued in Chap. 2, Mendel's concern was about the development of hybrids in their progeny, so it is historically flawed to understand Mendel's contribution as the introduction of the theory of heredity. In addition, this theory-based account overlooks the significance of the non-theoretical aspect of Mendel's contribution. For example, Mendel's experimental procedure that planting peas both in the field and in the pots indoor to protect plants from the foreign contamination was adopted by Tschemak (1900, 233).

Alternatively, based on our best historiography, one might regard Mendel's contribution to the study of heredity as the proposal of a theory of hybrid development. From a historical viewpoint, this account is better than the standard textbook interpretation. Mendel's concern was about hybrid development, and his work did make a substantial contribution to the study of hybrids (Gärtner 1849). However, there are still two problems. First, it is not obvious in what sense Mendel's theory of hybrid development was a contribution to the study of heredity. It is not straightforward to see why a theory of hybrid development was important for the study of heredity. Second, as I have repeatedly emphasised, a theory-based analysis of Mendel's work overlooks the non-theoretical contribution made by Mendel. Thus, it is still inappropriate to argue that Mendel's contribution to the study of heredity was the theory of hybrid development.

Another popular account of Mendel's contribution is that he made some important observations on the pattern of hybrid development, or a discovery of some important facts of hybrid development (e.g. the 1: 2: 1 ratio in the F_2 generation). However, the facts can never be discovered without a proper representation. As Bowler (1989, 6) puts it, "Facts only appear as facts within an appropriate conceptual scheme." Such an understanding overlooks Mendel's conceptual contribution to the origin of genetics by introducing new concepts (e.g. dominance) to describe and explain the phenomena. As Curt Stein indicates,

> [Mendel's paper "Experiments on Plant Hybrids"] does not simply announce the discovery of important facts by new methods of observation and experiment. Rather is an act of highest creativity, it presents these facts in a conceptual scheme which gives them general meaning. (Stern and Sherwood 1966, v)

Thus, Mendel's contribution was more than the discovery of some facts of hybrid development. Therefore, I suggest that a proper interpretation of Mendel's contribution should encompass both the theoretical and non-theoretical aspects of Mendel's work.

7.3.2 *The Exemplar-Based Analysis*

Following the exemplar-based analysis in Chaps. 5 and 6, I argue that Mendel's contribution to the study of heredity was that he provided an exemplary practice of hybrid development. As I have shown, Mendel introduced a set of contextually well-defined research problems on hybrid development and the corresponding solutions, and some components of his exemplary practice greatly inspired and influenced de Vries', Correns', Tschermak', and William Bateson's work, and lay down the cornerstone for the study of heredity in the twentieth century.

Compared with the traditional accounts, the exemplar-based account is better in two aspects. Firstly, it fits better with our best historiography of the origin of genetics. From a historiographical point of view, Mendel made four contributions: the novel approach to the study of the problem of development by focusing on the paired traits in the successive generations; the novel use of the terms "dominant" and "recessive" to conceptualise the paired traits and analyse the statistical relation of them; the proposal of the law of composition of hybrid fertilising cells; and the mathematical approach. As I have argued in Chap. 5, all these four contributions can be well explained in the exemplar-based analysis. The focus on paired traits can be illustrated as the problem-specification (MP1 $\rightarrow$ MP2) in Mendel's exemplary practice. The introduction of the conceptions of dominance and recessiveness can be characterised an exemplary way of conceptualising the phenomena. The proposal of the law of composition of hybrid fertilising cells (MH3) is an exemplary hypothesisation to account for the mechanism of the transmission of morphological traits, while Mendel's distinctive mathematical approach can be analysed a novel way of reformulating the hypotheses symbolically. Thus, the historiography of Mendel's contribution can be well explained in an exemplar-based framework.

Second, the exemplar-based account highlights both the theoretical and non-theoretical contribution made by Mendel's work. The traditional theory-based interpretations usually overlook the significance of the non-theoretical aspects of Mendel's work, while the exemplar-based account provides a fuller analysis of Mendel's legacy, especially in problem-defining and experimentation. Therefore, I argue that Mendel's contribution should be characterised as the introduction of contextually well-defined research problems on hybrid development and their corresponding solutions. Accordingly, in the next section, I shall propose a new explanation of the long neglect by accounting for why Mendel's contribution was neglected.

7.4 Why Mendel's Contribution Was Neglected

7.4.1 The Nature of the Long Neglect

Let us re-examine the problem of the long neglect. The problem originated from the "rediscovery" event in 1900. Conceiving the promising future of a Mendelian approach to the study of heredity, Bateson was puzzled by a fact that the significance of Mendel's work for the study of heredity was forgotten for 35 years.

> It may seem surprising that a work of such importance should so long have failed to find recognition and to become current in the world science. (Bateson 1902, 37)

It seemed natural for Bateson, as a leading proponent of Mendelism, to have such a puzzle. Given that Mendel's work was so important to the study of heredity, how could it be ignored or neglected for such a long time? Why was its significance not recognised earlier? However, this also reflects another problem. From the very beginning, the problem of the long neglect was deeply rooted in an unexamined understanding of Mendel's work: There was an obvious implication of Mendel's work for the problem of heredity. If so, it must be puzzling why so many leading naturalists in the field like Darwin and Francis Galton who made efforts to understand the phenomena and mechanism of heredity failed to recognise the valuable implication of Mendel's work. This is also why Bateson's counterfactual fantasy had been so widespread among historians and biologists for a long time.

> Had Mendel's work come into the hands of Darwin, it is not too much to say that the history of the development of evolutionary philosophy would have been very different from that which we have witnessed. (Bateson 1902, 39)

Such an analysis of the problem of the long neglect is clearly flawed. Remember that Mendel's concern was the development of hybrids in their progeny. It is not surprising that his work was not warmly received by those who studied heredity in the late nineteenth century. *What seems obvious to us was not obvious to Mendel's contemporaries, especially those who were particularly interested in the problem of heredity.* For most of Mendel's contemporaries, a paper on hybrid development would not obviously make a potential contribution to the study of heredity. In other words, the usefulness of Mendel's exemplary practice was not that obvious. The silence on the significance of Mendel's work before 1900 and the struggle and confusion of understanding Mendel's work by de Vries, Correns, Tschermak, and many others in the first few years of the twentieth century confirm this point. This well reflects that Mendel's exemplary practice on hybrid development did not seem to be an obvious supplement to the study of heredity in the nineteenth century.

7.4.2 The Exemplar-based Explanation

I have argued in Chaps. 5 and 6 that usefulness is an essential characteristic of a good exemplary practice. Accordingly, I propose that the introduction of Mendel's work to the study of heredity in 1900 can be characterised as the acceptance of the usefulness of Mendel's exemplary practice.

In the period 1900–1902, de Vries, Correns, Tschermak, and Bateson all acknowledged the usefulness of Mendel's exemplary practice to other research problems.

> [The law of hybrids as discovered by Mendel for peas] has a basic significance for the study of the units of which the species character is composed. (de Vries 1900a, 90, 1966, 117)

> In order to explain the facts, one must assume (as did Mendel) that ... (Correns 1900, 163, 1966, 125)

> [Mendel's principle] proves to be of the highest significance for the study of inheritance in general. (Tschermak 1900, 235, 1950, 44)

> [B]y the application of those [Mendelian] principles we are enabled to reach and deal in a comprehensive manner with phenomena of a fundamental nature, lying at the very root of all conceptions not merely of the physiology of reproduction and heredity, but even of the essential nature of living organisms, ... (Bateson 1902, 35)

More precisely, as shown in Sect. 5.5, the focus on a pair of differing traits (problem-defining), the concepts of dominance and recessiveness and their statistical relation (conceptualisation), the hypothesis of the morphological-cellular correspondence (hypothesisation), and his mathematical approach (conceptualisation and hypothesisation) are four constituents of Mendel's exemplary practice which were particularly useful for de Vries' and Bateson's' work on heredity, despite the lack of generality of the Mendelian laws. In contrast, none of these constituents of Mendel's exemplary practice was taken into serious consideration in the nineteenth century study of heredity. This is well corroborated by a traditional explanation that the long neglect of Mendel's work was due to the obscurity of his mathematical approach. In other words, the failure of recognising the usefulness of Mendel's symbolic representation is a key factor of the long neglect.

In addition, all de Vries, Correns, Tschermak, and Bateson repeated Mendel's exemplary practice on *Pisum* in their own ways. De Vries mainly repeated Mendel's problem-specification, conceptualisation, and hypothesisation to reinterpret his old data, drawn from his massive experiments on hybridisation in the 1890s. In particular, following Mendel's use of the concepts of dominance and recessive, de Vries (1900a, 1900b, 1900c) was able to re-analyse the data from his hybridisation experiments and to propose the law of segregation. Correns (1900), to a great extent, repeated most of Mendel's experiments and hypothesisation, as he aimed to test Mendel's work on hybrids. As I also mentioned, Tschermak repeated some of Mendel's procedure of experimentation. Bateson (1902) applied Mendel's statistical approach to the general problem of heredity. What is more, all of de Vries, Correns, Tschermak, and Bateson found Mendel's statistical analysis (e.g. the 3: 1 ratio) repeatable both conceptually and practically.

Thus, it is not hasty to conclude that the usefulness of Mendel's exemplary practice was appreciated by de Vries, Correns, Tschermak, and Bateson at the turn of the twentieth century. To some extent, it also explains why the significance of Mendel's work was well recognised in 1900 by these men. The failure of recognising the usefulness of Mendel's exemplary practice is one important intellectual explanation of why so many Mendel's contemporaries overlooked the significance of Mendel's work.

I also wish to highlight that this exemplar-based explanation is further corroborated by the Mendelian-Biometrician controversy in the first decade of the twentieth century. The Mendelians, led by Bateson, were optimistic on the future of the Mendelian approach to the study of heredity, while the Biometricians, led by Karl Pearson and W. F. R. Weldon, doubted the usefulness of Mendel's approach (especially his conceptualisation, hypothesisation, and experimentation) to study the phenomena of heredity. One main reason for Weldon (1902a, 1902b) to resist the Mendelian approach to the study of heredity was that he failed to repeat Mendel's way of distinguishing the dominant from recessive characters. In other words, for Weldon. Mendel's work failed to provide a repeatable way of conceptualisation. In contrast, Bateson was convinced by the usefulness of Mendel's exemplary practice. In particular, he maintained that Mendel's work provided a reliable framework for further investigation across different areas.

> [T]hose who as evolutionists or sociologists are striving for wider views of the past or of the future of living things may by the use of Mendelian analysis attain to a new and as yet limitless horizon. (Bateson 1909, 17)

> No one who is acquainted with Mendelian method will doubt that by its use practical breeders of animals and plans may benefit. (Bateson 1909, 291)

Moreover, Bateson (1909, 7) argued that by denying "the truth of Mendelian facts", the Biometricians were "delaying recognition of the value of Mendelism." Therefore, I argue that the Mendelian-Biometrician controversy reflects a fundamental disagreement on the usefulness of Mendel's exemplary practice. To a great extent, both the neglect and scepticism of the significance of Mendel's work for the study of heredity resulted from the failure of the recognition of the usefulness of Mendel's work.

7.4.3 *The Exemplar-Based Explanation and Old Intellectual Explanations*

Finally, I would like to remark that my exemplar-based explanation is not a comprehensive account covering all the facets of the problem of the long neglect. It is clear that there are many significant sociological factors involved that are worthy of further examination. These sociological factors are mutually interacted with the intellectual factors. However, what I aim at in this chapter is an intellectual account of why Mendel's contemporaries overlooked the significance of Mendel's work in the study of heredity, from a philosophical viewpoint. I find such a philosophical analysis as necessary and significant as a historico-sociological analysis. Moreover, the exemplar-based explanation provides a new way to reinterpret some traditional explanations. The problem of novelty and the prematurity of Mendel's idea can obviously be characterised as the failure of the recognition of the usefulness of Mendel's exemplary practice. The lack of generality can also be characterised as the failure of the repeatability of Mendel's approach. The obscurity of Mendel's mathematical approach and the problem of irrelevance can also be taken as the cases of how Mendel's contemporaries failed to appreciate the usefulness of Mendel's exemplary practice. Furthermore, the exemplar-based explanation can well account for the problems, which are unsolved by the traditional explanations. The problems of irrelevance, of novelty, of lack of generality, and of obscurity remained as serious problems at the turn of the twentieth century. The reason why these problems did not prevent de Vries, Correns, Tschermak, and Bateson from accepting Mendel's work is that all of them were able to recognise and appreciate the usefulness of Mendel's exemplary practice. Thus, I argue that the exemplar-based account provides a good framework to explain the intellectual nature of the problem of long neglect in two ways: It reinterprets some traditional explanations and well accounts for the unsolved puzzles.

7.5 Conclusion

In this chapter, I have shown that the exemplar-based analysis of the origin of genetics is helpful to explain the new version of the problem of the long neglect. The long neglect of the significance of Mendel's work for the study of heredity is explained in the way that the usefulness of Mendel's exemplary practice was not appreciated or recognised by those who studied the problem of heredity in the nineteenth century. The emergence of Mendelism in the first decade of the twentieth century arose with a gradual acceptance of the usefulness of Mendel's exemplary practice by de Vries, Correns, and especially Bateson and his collaborators, while both the long neglect of the significance of Mendel's work for the study of heredity in the nineteenth century and the resistance to Mendelism in the 1900s can be well characterised as the failure of the recognition of the usefulness of Mendel's exemplary practice.

References

Atherly, Alan, Jack Girton, and John McDonald. 1999. *The Science of Genetics*. Fort Worth, TX: Saunders College Publishing.

Bateson, William. 1902. *Mendel's Principles of Heredity: A Defence*. Cambridge: Cambridge University Press.

———. 1909. *Mendel's Principles of Heredity*. Cambridge: Cambridge University Press.

Blomberg, C.A. 1872. *Om Hybridbildning Hos de Faneroganma Vaxterna*. Stockholm University.

Bowler, Peter J. 1989. *The Mendelian Revolution: The Emergence of Hereditarian Concepts in Modern Science and Society*. London: The Athlone Press.

Brown, T.A. 1989. *Genetics: A Molecular Approach*. London: Chapman & Hall.

Callender, L.A. 1988. Gregor Mendel: An Opponent of Descent with Modification. *History of Science* 26 (1): 41–75.

Corcos, Alain F., and Floyd V. Monaghan. 1985. Role of de Vries in the Recovery of Mendel's Work I. Was de Vries Really an Independent Discoverer of Mendel? *Journal of Heredity* 76 (3): 187–190.

———. 1987a. Correns, an Independent Discoverer of Mendelism? I. An Historical/Critical Note. *Journal of Heredity* 78 (5): 330.

———. 1987b. Correns, an Independent Discovery of Mendelism? II. Was Correns a Real Interpreter of Mendel's Paper? *Journal of Heredity* 78 (6): 404–405.

Correns, Carl. 1900. G. Mendels Regel über das Verhalten der Nachkommenschaft der Rassenbastarde. *Berichte der Deutschen Botanischen Gesellschaft* 18 (4): 158–168.

———. 1966. G. Mendel's Law Concerning the Behavior of Progeny of Varietal Hybrids. In *The Origin of Genetics: A Mendel Source Book*, ed. Curt Stern and Eva R. Sherwood, translated by Leonie Kellen Piternick, 119–132. San Francisco, CA: W. H. Freeman and Company.

Dodson, Edward. 1955. Mendel and the Rediscovery of His Work. *The Scientific Monthly* 81 (4): 187–195.

Dorsey, M.J. 1944. Appearance of Mendel's Paper in American Libraries. *Science* 99 (2567): 199–200.

East, E.M. 1923. Mendel and his Contemporaries. *The Scientific Monthly* 16 (3): 225–237.

Gärtner, Carl Friedrich. 1849. *Versuche und Beobachtungen über die Bastarderzeugung im Pflanzenreich*. Stuttgart: Gedruckt bei K. F. Hering & Comp.

Garvey, William D., and Belver Griffith. 1971. Scientific Communication: Its Role in the Conduct of Research and Creation of Knowledge. *American Psychologist* 26 (4): 349–362.

Gasking, Elizabeth B. 1959. Why Was Mendel's Work Ignored? *Journal of the History of Ideas* 20 (1): 60–84.

Glass, Bentley. 1974. The Long Neglect of Genetic Discoveries and the Criterion of Prematurity. *Journal of the History of Biology* 7 (1): 101–110.

Hartl, Daniel L., David Freifelder, and Leon Snyder. 1989. *Basic Genetics*. Boston, MA: Jones and Barlett Publishers.
Hartwell, Leland, Leroy Hood, Michael Goldberg, Ann Reynolds, Silver Lee, and Ruth Veres. 2008. *Genetics: From Genes to Genomes*. Boston, MA: McGraw Hill Higher Education.
Harwood, Jonathan. 2000. The Rediscovery of Mendelism in Agricultural Context: Erich von Tschermak as Plant-Breeder. *Comptes Rendus de l'Academie Des Sciences - Serie III* 323 (12): 1061–1067.
Iltis, Hugo. 1932. *Life of Mendel*. Translated by Eden Paul and Cedar Paul. London: George Allen & Unwin Ltd.
Kottler, Malcolm J. 1979. Hugo de Vries and the Rediscovery of Mendel's Laws. *Annals of Science* 36 (5): 517–538.
MacRoberts, Michael H. 1985. Neglected or Unknown? *Annals of Science* 42 (3): 339–345.
Mayr, Ernst. 1982. *The Growth of Biological Thought*. Cambridge, MA: The Belknap Press.
Mendel, Gregor. 1866. "Versuche über Pflanzenhybriden." *Verhandlungen des Naturforschenden Vereins Brünn* IV (1865) (Abhandlungen): 3–47.
Monaghan, Floyd V., and Alain F. Corcos. 1986. Tschermak: A Non-Discoverer of Mendelism, I. An Historical Note. *Journal of Heredity* 77 (6): 468–469.
———. 1987. Tschermak: A Non-Discoverer of Mendelism. II. A Critique. *The Journal of Heredity* 78 (3): 208–210.
Olby, Robert Cecil. 1985. *Origins of Mendelism*. 2nd ed. Chicago, IL: The University of Chicago Press.
Orel, Vitězslav. 1996. *Gregor Mendel: The First Geneticist*. Oxford: Oxford University Press.
Posner, E., and J. Skutil. 1968. The Great Neglect: The Fate of Mendel's Classic Paper between 1865 and 1900. *Medical History* 12 (2): 122–136.
Punnett, Reginald Crundall. 1919. *Mendelism*. 5th ed. London: Macmillan and Co., Limited.
Sandler, Iris, and Laurence Sandler. 1985. A Conceptual Ambiguity That Contributed to the Neglect of Mendel's Paper. *History and Philosophy of the Life Sciences* 7 (1): 3–70.
Schmalhausen, I.F. 1874. *O Rastitelnych Pomesjach-Nabljudenija Iz Petersburskoy Flory*. St. Petersburg University.
Simunek, Michal, Uwe Hoßfeld, and Olaf Breidbach. 2011. 'Rediscovery Revised' – The Coopertation of Erich and Armin von Tschermak-Seysenegg in the Context of the 'Rediscovery' of Mendel's Law in 1899-1901. *Plant Biology* 13 (6): 835–841.
Sorsby, Arnold. 1965. Gregor Mendel. *British Medical Journal* 1 (5431): 333–338.
Stent, G.S. 1972. Prematurity and Uniqueness in Scientific Discovery. *Scientific American* 227 (6): 84–93.
Stern, Curt, and Eva R. Sherwood, eds. 1966. *The Origin of Genetics: A Mendel Source Book*. San Francisco, CA: W. H. Freeman and Company.
Stubbe, Hans. 1972. *History of Genetics from Prehistoric Times to the Rediscovery of Mendel's Laws*. Cambridge, MA: The MIT Press.
Suzuki, David, Anthony Griffiths, Jeffrey Miller, and Richard Lewontin. 1981. *An Introduction to Genetic Analysis*. New York: W.H. Freeman and Company.
von Tschermak, Erich. 1900. Über Künstliche Kreuzung bei Pisum Sativum. *Berichte der Deutschen Botanischen Gesellschaft* 18 (6): 232–239.
———. 1950. Concerning Artifical Crossing in Pisum Sativum. Translated by Aloha Hannah. *Genetics* 35 (5 (2)): 42–47.
———. 1951. The Rediscovery of Gregor Mendel's Work: An Historical Retrospect. *Journal of Heredity* 42 (4): 163–172.
de Vries, Hugo. 1900a. Das Spaltungsgesetz der Bastarde (Vorlaufige Mittheilung). *Berichte der Deutschen Botanischen Gesellschaft* 18 (3): 83–90.
———. 1900b. Sur la Loi de Disjonction des Hybrides. *Comptes Rendus de l'Academie des Sciences (Paris)* 130: 845–847.
———. 1900c. Sur les Unités des Caractères Spécifiques et Leur Application à l'étude des Hybrides. *Revue Génértate de Botanique* 12: 257–271.

———. 1966. The Law of Segregation of Hybrids. In *The Origin of Genetics: A Mendel Source Book*, edited by Curt Stern and Eva R. Sherwood, translated by Evelyn Stern, 107–117. San Francisco, CA: W. H. Freeman and Company.

Weiling, Franz. 1991. Historical Study: Johann Gregor Mendel 1822-1884. *American Journal of Medical Genetics* 40 (1): 1–25.

Weinstein, Alexander. 1962. The Reception of Mendel's Paper by His Contemporaries. *Proceedings of the Tenth Congress of the History of Science* 2: 997–1001.

Weldon, Walter Frank Rapheal. 1902a. Mendel's Laws of Alternative Inheritance in Peas. *Biometrika* 1 (2): 228–254.

———. 1902b. On the Ambiguity of Mendel's Categories. *Biometrika* 2 (1): 44–55.

Wilkie, J.S. 1962. Some Reasons for the Rediscovery and Appreciation of Mendel's Work in the First Years of the Present Century. *The British Journal for the History of Science* 1 (1): 5–17.

Zirkle, Conway. 1964. Some Oddities in the Delayed Discovery of Mendelism. *Journal of Heredity* 55 (2): 65–72.

———. 1968. The Role of Liberty Hyde Bailey and Hugo de Vries in the Rediscovery of Mendelism. *Journal of the History of Biology* 1 (2): 205–218.

Part III
Philosophy

In Part II, I proposed a new integrated HPS method, the exemplar-based approach, to analyse and interpret the development and the progress in early genetics. I have argued that the progress in the history of genetics could be well characterised as the increase of the usefulness of problem-defining and problem-solving, and the significance of Mendel's work for the study of heredity was neglected because its usefulness was not appreciated. In this part, I argue that the exemplar-based approach sheds new light on some issues of general philosophy of science. In Chap. 8, I show that putting conceptual practice (or conceptualisation) in the broader context of scientific practice (i.e. exemplary practice) motivates a new way to analyse and account for conceptual change and continuity in science. In Chap. 9, I show that the gap problem of H-D evidence can be resolved if philosophers shift attention from evidence to evidential practice. In Chap. 10, I propose a new criterion of theory choice in science, promisingness, inspired by the notion of usefulness. All these arguments are illustrated by the case studies from the history of genetics.

Chapter 8
A New Mode of Conceptual Continuity

Abstract When we say that a scientific concept *C* changes, we mean that (1) at least some aspect or component of *C* varies, and (2) there is still something continuous in the historical development of *C*. However, how to account for the continuity in a radical conceptual change has been a difficulty. This chapter overcomes this difficulty by providing a new account of conceptual continuity in science. I begin with an examination of the historical development of the concept of dominance and highlight the inadequacies of the analytic, the cognitive, and the practice-based approaches. Then I introduce a holistic approach to conceptual change. By distinguishing two modes of continuity, I also propose a new mode of conceptual continuity. Finally, I illustrate and defend my approach by revisiting my historical case study.

Keywords Conceptual continuity · Conceptual change · Conceptual practice · Holistic continuity · Scientific practice · Dominance

8.1 Conceptual Change: Variance and Continuity

Conceptual change in science has two distinctive features. One is, obviously, variance. When we say that a scientific concept *C* changes, we mean that at least some aspect or component of *C* varies. The other is continuity. Despite the change of its aspects or components, there is still something continuous in the historical development of *C*. Thus, an adequate philosophical analysis of conceptual change in science should well characterise and account for not only in what way a scientific concept varies, but also in what sense its practice[1] continues. In general, there are three approaches to conceptual change: the analytic approach, the cognitive approach, and the practice-based approach. The analytic approach accounts for the

---

[1]In this chapter, I use "the practice of a concept" and "the use of a concept (in practice)" interchangeably.

Y. Shan, *Doing Integrated History and Philosophy of Science: A Case Study of the Origin of Genetics*, Boston Studies in the Philosophy and History of Science 320,
https://doi.org/10.1007/978-3-030-50617-9_8

continuity by arguing that the reference of a concept is invariant when its meaning changes (e.g. Putnam 1975; Sankey 1994; Psillos 1999). As Hilary Putnam (1973, 175) illustrates in the case of "electricity", we have different meanings of the concept of electricity at different times in the history of physics, but its reference remains stable. The meaning variance shows the physicists' attempts to explain and understand the phenomena related to "electricity", which capture the continuity of conceptual practice. Thus, according to the analytic approach, the continuity is explainable in terms of reference. (Referential continuity).

The cognitive approach (e.g. Thagard 1990, 1992; Barsalou 1992; Andersen and Nersessian 2000; Votsis and Schurz 2012) construes a scientific concept as a frame-like structure, which consists of a recursive system of attributes, each of which takes a range of values. Accordingly, conceptual change is characterised as the change of the attributes or of the values. Thus, the continuity of conceptual practice can be accounted for in the way that there are only some changes of the attributes of a stable conceptual structure. (Structural continuity) For example, Hanne Andersen and Nancy Nersessian (2000) argue that despite the change in some attributes (e.g. ontological status), "there is still significant continuity between the Maxwellian and the Einsteinian concepts of field." As illustrated in Fig. 8.1, the concept of electromagnetic field as a structure remains stable, especially given that two key attributes (i.e. function and causal power) are invariant.

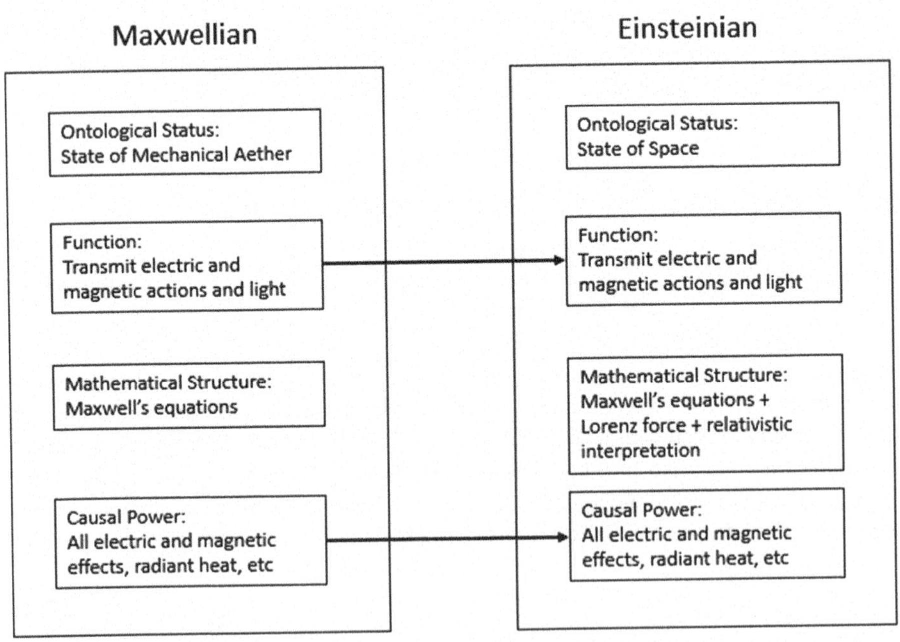

Fig. 8.1 The evolution of the electromagnetic field concept

The practice-based approach (e.g. Nersessian 2008; Brigandt 2010, 2012; Feest 2010; Chang 2011; Boon 2012; Kindi 2012; Waters 2014) highlights the significance of the practical function of a scientific concept and accounts for the continuity in conceptual change in terms of practical function. (Practical continuity) Ingo Brigandt (2010), for example, argues that a scientific concept consists of its meaning, its reference, and its epistemic goal pursued by its use. Accordingly, the continuity of conceptual practice is defended as long as there is a stable epistemic goal (and use) of a scientific concept, no matter how its meaning or reference varies. In a similar vein, Uljana Feest (2010) proposes that scientific concepts are tools to enable experimental interventions in the process of investigating the purported object, and thus argues that the continuity of the practices of a concept C is guaranteed if the purported object under investigation remains unchanged.

However, none of the analytic, the cognitive, and the practice-based approaches exhaustively captures all types of conceptual continuity. Clearly, the analytic approach fails to account for the continuity in a conceptual change where both meaning and reference vary. The cognitive approach is inadequate to characterise the continuity in a radical change where there is a fundamental structural variance. Nor does the practice-based approach provide a good account of the continuity in a conceptual change in which the practical function changes. It is not unusual for a scientific concept to vary its meaning, reference, and practical function radically in history. The concept of the gene is a clear case.[2] An important lesson drawn from the historical development of the concept of the gene is that, as Raphael Falk (1986, 170) summarises, "[T]he term 'gene' has changed both its meaning and its reference." Therefore, referential invariance is inapplicable to account for the continuity in the case of the gene concept. Similarly, it is extremely difficult to characterise the change of the concept of the gene as merely the changes of the attributes or values of a stable structure. In other words, the continuity in the history of the gene concept is not easily accounted for in terms of the change of the attributes. The gene concept used by early Mendelians and the gene concept used by molecular biologists cannot be captured in the same structure. The practice-based approach is also inadequate. As Brigandt (2010, 27) himself even recognises, both the semantic and practical aspects of a scientific concept may change. For example, the epistemic goal of the concept of the gene has been changed again and again. The classical geneticists in the 1920s and 1930s used the concept of the gene to explain patterns of inheritance, while the molecular biologists' use in the 1960s and 1970s aimed at accounting for the molecular function of genes. The practical function of the gene concept is not stable, either. That said, *there is a strong historical connection* between the variants of the gene concept. Thus, how to account for the continuity in a radical conceptual

[2]For a detailed study of the historical development of the gene concept, see Peter Portin (1993) and Paul Griffiths and Karola Stotz (2007).

change, in which all of the components change (e.g. the gene concept), poses a serious difficulty.[3]

This chapter provides a new account of conceptual continuity in order to overcome this difficulty. In Sect. 8.2, I illustrate the problems of old approaches with a historical analysis of the development of the concept of dominance. In Sect. 8.3, I propose a holistic approach to conceptual change. In Sect. 8.4, by distinguishing two modes of continuity, I also propose a new mode of conceptual continuity. In Sect. 8.5, I illustrate and defend this mode by revisiting my historical case study.

8.2 What if Everything Changes? The Case of the Concept of Dominance

John Goss read a paper in the meeting of the Horticultural Society of London in 1822, which reported an interesting phenomenon of transmission of morphological traits derived from his experiments on peas.

> In the summer of 1820, I deprived some blossoms of the *Prolific blue* of their stamina, and the next day applied the pollen of a *dwarf Pea,* and of which impregnation I obtained three pods of seeds. In the following spring, when these were opened, in order to sow the seed, I found, to my great surprise, that the colour of the Peas, instead of being a deep *blue,* like their female parent, was of a *yellowish white,* like the male. (Goss 1824, 234–35)

The similar remarkable phenomena in pea hybridisation was also observed and recorded by Thomas Andrew Knight (1799, 1824) and Alexander Seton (1824) in England. However, Gregor Mendel (1866) was the first to conceptualise the phenomenon in terms of "dominance" and "recessiveness." It should be noted that Mendel was not the first to use the term "dominance" in the nineteenth century to characterise the phenomena in the crossing experiments. Count Giorgio Gallesio and Michel Sageret already used the terms "dominante" and "dominait" respectively a few decades before Mendel.

> I fecundated some carnations having white flowers with the pollen taken from carnations bearing red flowers, and conversely; the seeds I gathered from them gave me carnations with mixed colors ... It follows that the crossing of these, not being natural, produces varying effects which bear the imprint of their different sources in the proportion to which a given one is dominant (*dominante*). (Gallesio 1816, 76, 79; Stubbe 1972, 107–8)

> The acid flavor of the muskmelon is encountered in the form of the cantaloupe and the snake-melon; in others, the form of the cantaloupe dominated (*dominait*). (Sageret 1826, 307; Roberts 1929, 121)

[3]Some may question this motivation by arguing that a radical conceptual change may simply suggest a concept-shift. If so, it is not necessary to explore a new account of conceptual continuity. In response, I argue that the failure to account for the continuity in a radical conceptual change does not eliminate the possibility that there is an overlooked mode of conceptual continuity. This is what I aim to explore in this chapter. Moreover, I contend that my new mode of conceptual will suggest a new distinction between radical conceptual change and concept-shift.

Nevertheless, neither Gallesio nor Sageret defined the concept of dominance in an explicit way as Mendel did.[4] In the most cases, Mendel used the term "dominance" to designate the morphological trait with a certain behaviour (dominance$_{\mathrm{Mmb}}$). Occasionally, Mendel also simply referred it to a morphological trait (dominance$_{\mathrm{Mm}}$). There was a subtle difference between Mendel's uses of dominance$_{\mathrm{Mmb}}$ and dominance$_{\mathrm{Mm}}$. Let me illustrate Mendel's concept of dominance with his hybridisation experiments. (See Fig. 2.1) When purely bred yellow pea seeds (A) are crossed with purely bred green pea seeds (a), all of the seeds obtained in the next generation (F1) are yellow (Aa). When these yellow seeds (Aa) are self-fertilised, then some of the seeds obtained in the next generation (F2) are yellow (A or Aa), while the others are green (a). For these seeds self-fertilised, all of the green seeds (a) produce green seeds (a), while some yellow seeds (A) produce yellow seeds (A) only, and other yellow seeds (Aa) produce both yellow (A or Aa) and green seeds (a). Thus, yellowness of the seeds is defined as dominant$_{\mathrm{Mm}}$, because it dominates throughout the generations compared with greenness of the seeds. However, there is some difference between the behaviours of yellow seeds in their progeny. Some yellow seeds (A) only produce yellow seeds, while other yellow seeds (Aa) produce both yellow and green seeds. The yellowness of the seeds which only produce yellow seeds is defined as dominant$_{\mathrm{Mmb}}$ (form), while the yellowness of the seeds which produce both yellow and green seeds is defined as hybrid (form).

Along with his work on hybrid development, Mendel's concept of dominance was to some extent neglected by those who were studying the problem of heredity until 1900. By introducing Mendel's work into the study of pangenesis, Hugo de Vries (1900a, 1900b, 1900c) proposed the law of segregation in terms of dominance/recessiveness.

> In the hybrid the simple differential character from one of the parents is accordingly visible or dominant while the antagonistic character is in the latent condition or recessive. (de Vries 1900b, 845, 1950, 30)

> Of the two antagonistic characters, Mendel calls the one visible in the hybrid the dominating, the latent one recessive. (de Vries 1900a, 85, 1966, 111)

As I have argued in Sect. 3.3.2, de Vries' usage of "dominant" (1900a, 1900b, 1900c) was genuinely different from Mendel's (1866). For de Vries, the terms "dominant" and "recessive" were used to label two different pairs of things. On the one hand, de Vries referred these to the pair of morphological traits (dominance$_{\mathrm{Dm}}$).

> The antagonistic characters [*caractères*] ordinarily remain combined during all of the vegetative life, one dominant, the other latent. (de Vries 1900b, 845, 1950, 30)

> The dominating and the recessive traits [*Merkmal*] are shown to be constant in the progeny, … In this experiment they yielded an average of 77% with the dominating and 23% the recessive trait [*Merkmal*]. (de Vries 1900a, 88, 1966, 114)

[4]It is not clear whether Mendel was familiar with Gallesio's or Sagaret's work. Nor is it clear whether Mendel's usage of the term "dominance" was influenced by theirs.

On the other hand, de Vries referred those to the pair of hereditary characteristics, or qualities (dominance$_{Dc}$).

> One can unify the whole of those results by supposing that the two antagonistic qualities [*qualities*], dominant and recessive, are distributed [mutually exclusively] in equal parts to the pollen just as to the ovules. (de Vries 1900b, 847)

> Of the two antagonistic [characteristics (*Eigenschaften*)], Mendel calls the one visible in the hybrid the dominating, the latent one recessive. (de Vries 1900a, 85, 1966, 111)

Neither *Eigenschaften* (*or qualities*) nor *Merkmal* (or *caractères*) can be simply conflated. The word *Merkmal*, also used by Mendel, generally referred to what nowadays we call the morphological trait, while *Eigenschaften* was originally used by de Vries in his book *Intracellular Pangenesis* (1889) to denote the hereditary property, which could be passed onto the next generation. It is clear that Mendel never used "dominant" to designate anything unobservable, say, hereditary property. In other words, de Vries' use of dominance$_{Dc}$ was novel.

Carl Correns (1900), another "rediscoverer" of Mendel's work, also differed from Mendel in using the terms "dominant" and "recessive". Although Correns also employed the terms "dominant" and "recessive" in two ways, his usage was slightly different. For Correns, the terms were attributed to both morphological traits (dominance$_{Cm}$) and anlagen (dominance$_{Ca}$).

> In many pairs one trait, or rather the anlage thereof, is so much stronger than the other trait, or its anlage, that the former alone appears in the hybrid plant, while the latter does not show up at all. This one may be called dominating, the other one the recessive, . . . (Correns 1900, 159, 1966, 121)

Correns' "anlagen" originated from August Weismann's terminology (1892), which referred to the primary constituents in the germ cell. Correns used it to implicitly designate the hereditary material for a morphological trait in the nuclei and the unit of segregation during the formation of the reproductive nuclei. Correns' dominance$_{Ca}$ was obviously different from de Vries' dominance$_{Dc}$ and Mendel's dominant$_{Mmb}$. De Vries' dominance$_{Dc}$ was about a property, while Correns' dominance$_{Ca}$ was about something physical.[5] In short, it was the first time in history that the term "dominant" was used to describe the hereditary material.

As the leading supporter of Mendelism, William Bateson (1902) used the terms "dominant" and "recessive" in a similar way to Correns'. On the one hand, the terms were applied to classify the morphological traits (dominance$_{Bm}$). On the other hand, the terms were applied to label the types of allelomorphs, or, unit-characters (dominance$_{Ba}$). Both Correns and Bateson attributed "dominant/recessive" to the observable traits and the hereditary material respectively, despite their different names for the unit of hereditary bearers. It should be highlighted that de Vries,

[5]For de Vries, there is a difference between hereditary property and hereditary material. Hereditary property is caused by hereditary material (that is, pangen, in his terminology). However, de Vries was never bold enough to postulate that dominant property was caused by a dominant pagen. This is the major difference between Correns' dominance$_{Ca}$ and from de Vries' dominance$_{Dc}$.

Table 8.1 The conceptions of dominance in early genetics

	Dominant/Recessive
Morphological traits	Mendel (1866), de Vries (1900a, 1900b, 1900c), Correns (1900), Tschermak (1900a, 1900b), Bateson (1902), Morgan (1926), Morgan et al. (1915)
Morphological traits with a certain behaviour in the progeny	Mendel (1866)
Hereditary characteristics	De Vries (1900a, 1900b, 1900c)
Hereditary material	Correns (1900), Bateson (1902), Morgan (1926), Morgan et al. (1915)

Correns, and Bateson all regarded the concept of dominance as the central part of the Mendelian theory of heredity. De Vries formulated the law of segregation in terms of dominance$_{\mathrm{Dm}}$ and dominance$_{\mathrm{Dc}}$. Correns' Mendelian rule (1900) was also implicitly based on dominance$_{\mathrm{Cm}}$ and dominance$_{\mathrm{Ca}}$. Bateson (1902, 26) explicitly included dominance$_{\mathrm{Bm}}$ and dominance$_{\mathrm{Ba}}$ in the seven new conceptions of heredity.

In the 1910s and 1920s, T. H. Morgan and his colleagues synthesised the Mendelian theory with the chromosome theory of heredity to establish a highly successful theory of inheritance. The terms "dominant" and "recessive" were still used from time to time in their work (e.g. Morgan et al. 1915; Morgan 1926) to classify the morphological traits and the hereditary factors.[6] However, the concept of dominance was not as important as it was in the 1900s. It was no longer a central concept of the study of heredity. As Lindley Darden (1991, 230) acutely points out, "[the claim on dominance/recessiveness] was not stated as a generalization in the theory of the gene." In other words, the theoretical significance of the concept of dominance was downplayed since the 1910s.

Thus, as summarised in Table 8.1, the meaning and the reference of the concept of dominance varied in the history of genetics. Thus, the analytic approach clearly fails to well account for the continuity of the practice of the concept of dominance, because it is extremely difficult to identify any referential continuity. Moreover, the continuity of the practice of the concept of dominance is not easily accounted for in terms of structures, either. As illustrated in Fig. 8.2, all of dominant$_{\mathrm{Mmb}}$, dominance$_{\mathrm{Dc}}$, and dominance$_{\mathrm{Dc}}$ are represented in different frames with different attributes and values. It is difficult to compare these frames, because the attributes and the values are so different in different frames. In other words, the continuity of the practice of the concept of dominance cannot be well characterised by the invariant value or attribute of the same frame-like structure.

In fact, not only the semantic aspect, but the practical aspect of the concept of dominance changed radically. Mendel's initial concern in his paper (1866) was to study the development of pea hybrids in their progeny. The concept of dominance was introduced at first to classify the pairs of the differing morphological traits by

[6]The hereditary factors were simply called factors (Morgan et al. 1915), while they were consistently named "genes" (Morgan 1926).

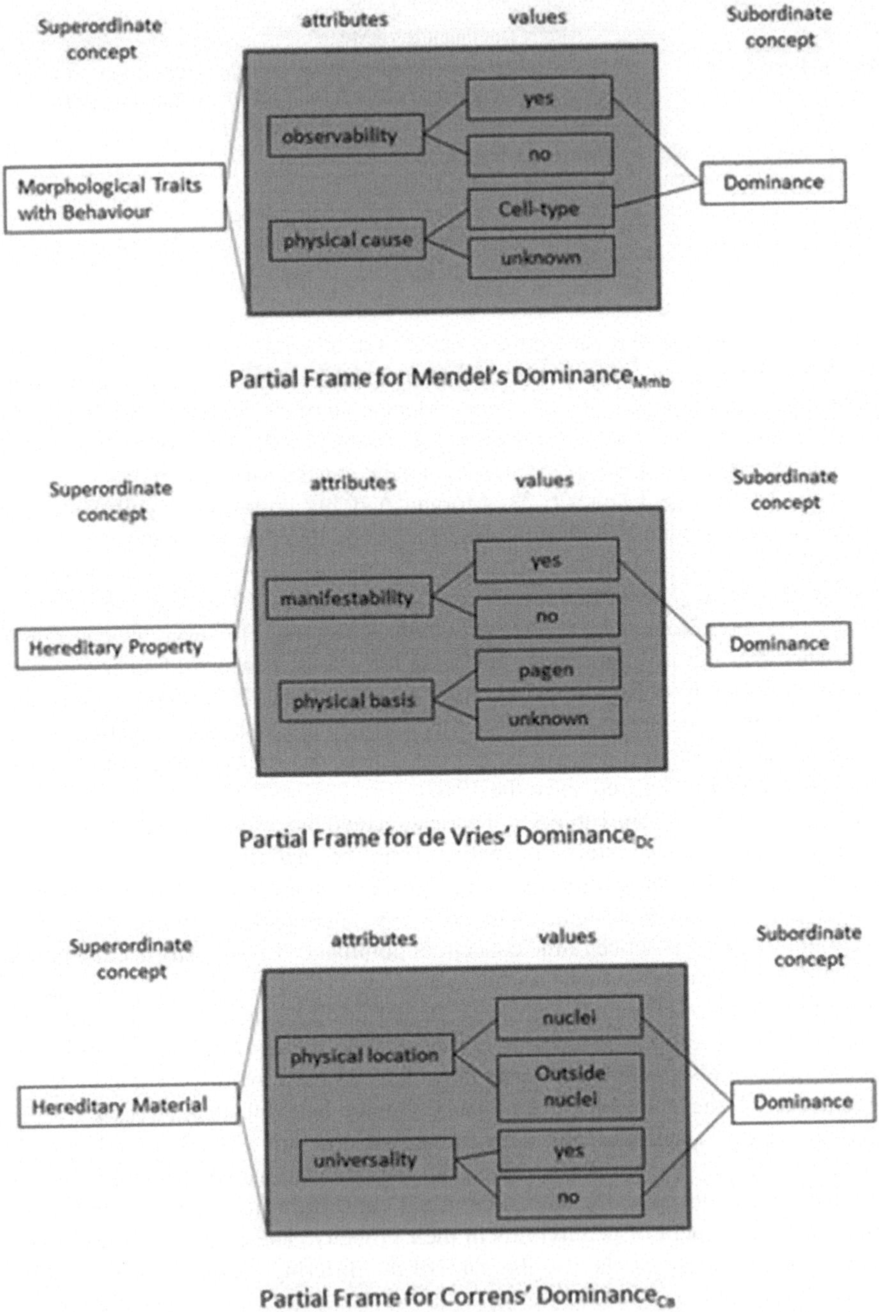

Fig. 8.2 Partial frames for the conceptions of dominance

observing their behaviour in the successive generations. As I have argued in Sect. 5.5, Mendel's dominant$_{Mmb}$ was important for his work in the sense that it lay down the conceptual foundation for his problem-specification (MP1 → MP2), analysis of data, summary of the statistical regularity (e.g. the 1: 2: 1 ratio), and proposal of the laws. In particular, the statistical analysis cannot be made meaningfully without his concept of dominance.

De Vries' initial concern (1900a, 1900b, 1900c) was to introduce Mendel's approach to test his theory of pangenesis (1889) experimentally (DP1). Mendel's concept of dominance was adopted with a modification, as it was useful to refine DP1 to a more specific problem DP2[7] and to formulate testable hypotheses (e.g. DH6 and DH7). As a result, de Vries' use of the concept of dominance (i.e. dominance$_{Dc}$) was an incorporation of Mendel's and his old concept of activeness. On the one hand, he followed Mendel's use of dominance to classify the morphological traits as a tool to re-analyse the data obtained in his hybridisation experiments. On the other hand, he attempted to map Mendel's "dominating/recessive" to his "active/latent" to describe the state of pagens, though he did not succeed eventually.

Correns' concern (1900) was to examine Mendel's work. So, it was not surprising that he tried to follow Mendel's practice, including conceptualisation, step by step. Nevertheless, his usage of the concept of dominance, especially dominance$_{Dc}$ was accommodated with the most update-to-date knowledge of cytology at the time.

Bateson (1902) aimed to develop a Mendelian approach to the study of heredity. Thus, his usage of the concept of dominance reflected an attempt to incorporate Mendel's terminology to the problem of heredity. Although de Vries, Correns, and Bateson differed in their research problems, all of their works were based on Mendel's approach. De Vries and Bateson were trying to incorporate Mendel's approach into their works, while Correns focused on testing Mendel's hypotheses. Therefore, it is natural that the concept of dominance lies in the core of their hypothesisation.

In contrast, Morgan and his colleagues were working on a general theory of inheritance which aimed at predicting and explaining the patterns of inheritance in a broader domain. The phenomena to be explained were much more complex than the transmission of the differing traits of plant hybrids. Neither Mendel's hypotheses nor his concepts were useful anymore. In particular, with the downplay of the significance of the concept of dominance, Morgan extended the explanatory domain of his theory. As Darden explicates,

> Deleting dominance from the set of theoretical components allowed Mendelians segregation (i.e. other theoretical components) to be extended to include characters that blended or showed new forms in heterozygous hybrids. Deletion of the empirical generalization about dominance from the domain allowed the scope of the domain to be expanded to include characters that did not show dominance. Thus, deletion of the theoretical component resulted

[7]In order to investigate DP1, de Vries (1900b, 845) began with the simplest case by examining the hybrid plants differing in a pair of antagonistic traits: "What is the change of a pair of antagonistic traits of hybrids?" (DP2)

in an expansion of the scope of the domain of the theory and the removal of an uncashed promissory note from the theoretical components. (Darden 1991, 231)

Therefore, the practical aspect of the concept of dominance changed from Mendel to Morgan significantly as well. More precisely, there was no stable epistemic goal of the concept of dominance in practice. Nor was any stable purported object under investigation of the concept of dominance in history. In short, the continuity of the practice of the concept of dominance cannot be well interpreted as a kind of practical continuity.

In summary, both of the semantic and practical aspects of the concept of dominance varied in the history of genetics. The historical development of the concept of dominance is highly suggestive. It provides a clear counterexample to all of the analytic, the cognitive, and the practice-based approaches to conceptual change by highlighting a difficulty.[8] These analyses of conceptual change begin with a constitutional analysis of scientific concepts. (A concept *C* consists of *X*, *Y*, *Z*, . . .) Then, the variance and the continuity are characterised in terms of constituents. (To examine whether *X*, *Y*, *Z*, . . . changes or not respectively.) However, as illustrated in the case of the concept of dominance, a fine-grained analysis of the history of sciences may provide counterexamples, no matter how a concept is constitutionally analysed. Therefore, a new account of conceptual continuity is desired in order to account for the continuity in a radical conceptual change in which all of the components of a concept change.

8.3 A Holistic Approach to Conceptual Change

There are some positive lessons from the case of the concept of dominance. It supports the rationale of the practice-based approach to conceptual change to a great extent. Scientific concepts are more than linguistic entities. As Stephen Toulmin (1972, 8) indicates, "[A scientific concept] has something to do with how we use language; or with how we structure our experience, or with how we categorize the objects we have to deal with . . . something . . ." In other words, we should not only examine what scientific concepts are from a semantic point of view, but also investigate how scientific concepts are used in practice. Accordingly, conceptual change should also be examined in terms of practices.

A related lesson is that the evolution of the concept of dominance highlights that conceptual practice is not an independent activity in scientific practice. For example,

[8]Some may wonder why the term "dominance" in different contexts (e.g. Mendel 1866; de Vries 1900b; Correns 1900; Bateson 1902; Morgan 1926) still refers to the same concept. Why cannot Mendel's "dominance" and de Vries' "dominance" be simply understood as different concepts? A brief explanation is that the uses of the term "dominance" in these contexts are historically linked. It is not a historical coincidence that Mendel and de Vries happened to use the same term. Rather, as I have argued in Chap. 3, de Vries's use of dominance was influenced by Mendel's use. There is a historical connection between the uses of the term "dominance."

Mendel's use of the concept of dominance is intertwined with his proposal of the laws and the design of the experiments. De Vries', Correns', Bateson's, and Morgan's uses of the concept of dominance are all well embedded in their research. Thus, a better understanding of conceptual practice and its relation to other scientific activities is necessary for a comprehensive analysis of conceptual change and continuity. As Nersessian (2008, 5) puts it, "Conceptual changes need to be understood in terms of the people who create and change their representations of nature and the practices they use to do so."

It has been recognised for a long time that conceptual practice is key to problem-solving in science (e.g. Kuhn 1970, 35–42; Nersessian 1984, 144–59, 2008, 2–18; Andersen and Nersessian 2000). Furthermore, as I have argued in Sect. 5.4, conceptual practice (or conceptualisation) not only contribute to problem-solving, but also to problem-defining. In other words, conceptual practice should not be analysed independently. It does Thus, I propose that *a scientific concept should be understood as a tool for scientists to define and solve the problems in practice*. More specifically, *a scientific concept is used as a tool to contribute to various activities in scientific practice, including problem-defining, hypothesisation, experimentation, and reasoning*. Accordingly, *we should understand conceptual practice and its change in a broader context of scientific practice*. In short, I suggest that we should take a holistic approach to conceptual change. When analysing conceptual change, we should not only scrutinise how the components of a scientific concept change, but also articulate how the change contributes to other activities in a broader context.

Basically, this holistic approach is still a practice-based approach. However, I wish to highlight that my holistic approach differs from the other practice-based approaches in the following ways. Firstly, compared with Brigandt's approach (2010) and Feest's approach (2010), the holistic approach provides a finer-grained analysis of the practical function of a scientific concept. Rather than roughly interpreting the practical function as the epistemic goal, I articulate the practical function or the use of a scientific concept as an instrumental tool to define and solve problems by means of problem-defining, hypothesisation, experimentation, and reasoning. Secondly, my approach highlights the significance of conceptual practice in the process of problem-defining, which was to some extent overlooked in the past. Most of the practice-based approaches (e.g. Brigandt 2010; Feest 2010) focus on how conceptual practice contributes to problem-solving. Nersessian (1984, 2008) even argues that conceptual change is a process of problem-solving. However, as illustrated in Mendel's use of the dominance concept, conceptual practice also contributes to problem-defining. In other words, conceptual practice not only helps to achive an epistemic goal, but also helps to set an epistemic goal. Thirdly, as I shall elaborate in the following two sections, my holistic approach helps to provide a new perspective to understand the continuity of conceptual practice, and a fuller understanding of the development of scientific practice in a broader context.

8.4 Two Modes of Continuity and Conceptual Continuity

In order to account for the continuity in the radical conceptual change, I argue that a new mode of continuity has to be introduced. The traditional examination of the continuity of conceptual practice is deeply rooted in a particular account of continuity: When talking of continuity, philosophers typically assume that there is some invariant retention of certain components throughout the change. As illustrated in Fig. 8.3, the green area is invariantly preserved from Stage 1 to Stage 3, so it seems natural to conclude that there is continuity from Stage 1 to Stage 3.

Correspondingly, when talking of the continuity of conceptual practice, philosophers take for granted that there must be some invariant retention of certain elements of the concept, whether semantic or practical, in history. Thus, in order to defend the continuity of conceptual practice, it is necessary to identify the invariant elements preserved throughout history. As illustrated in Fig. 8.4, the series of conceptual practices (from Practice 1 to Practice 3) is continuous, for there are some invariant elements of the concept preserved throughout the time. (Mode 1).

However, it should be noted that Mode A is not the only mode of continuity. There is another mode of continuity. Although there are no components invariantly retained throughout the change, there are some components temporarily retained in certain periods. As illustrated in Fig. 8.5, the green area is retained from Stage 1 to Stage 2, the blue area is retained from Stage 2 to Stage 3, and the red area is retained from Stage 3 to Stage 4. None of these areas are invariantly retained from Stage 1 to Stage 4. Nevetheless, it is undeniable that there is some kind of continuity from Stage 1 to Stage 4.

Fig. 8.3 Continuity mode A

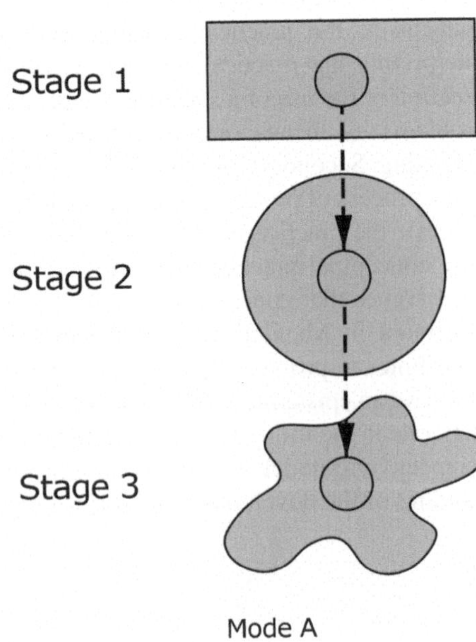

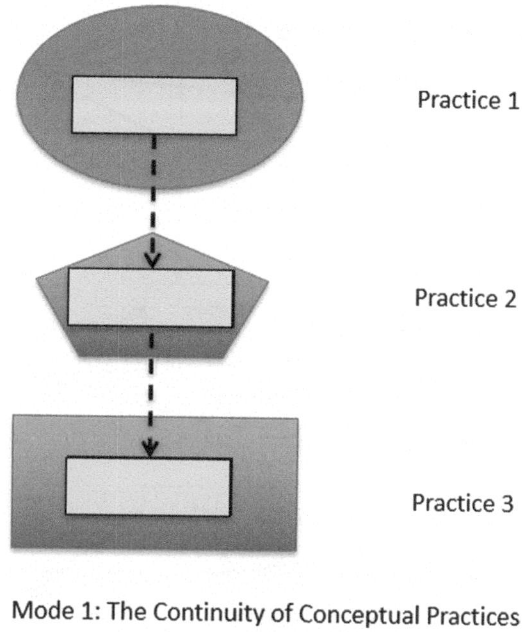

Fig. 8.4 Conceptual continuity mode 1

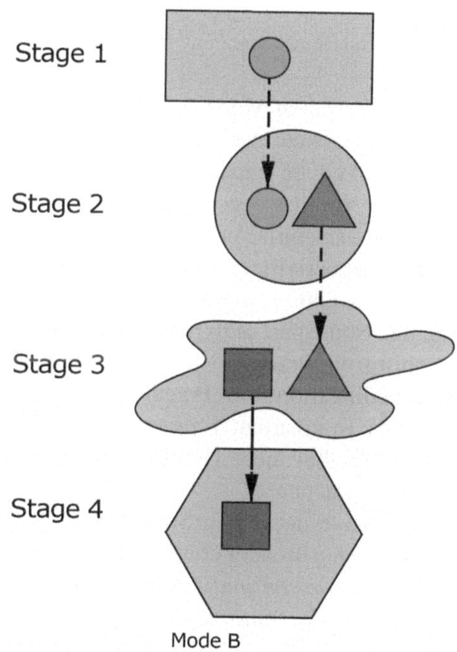

Fig. 8.5 Continuity mode B

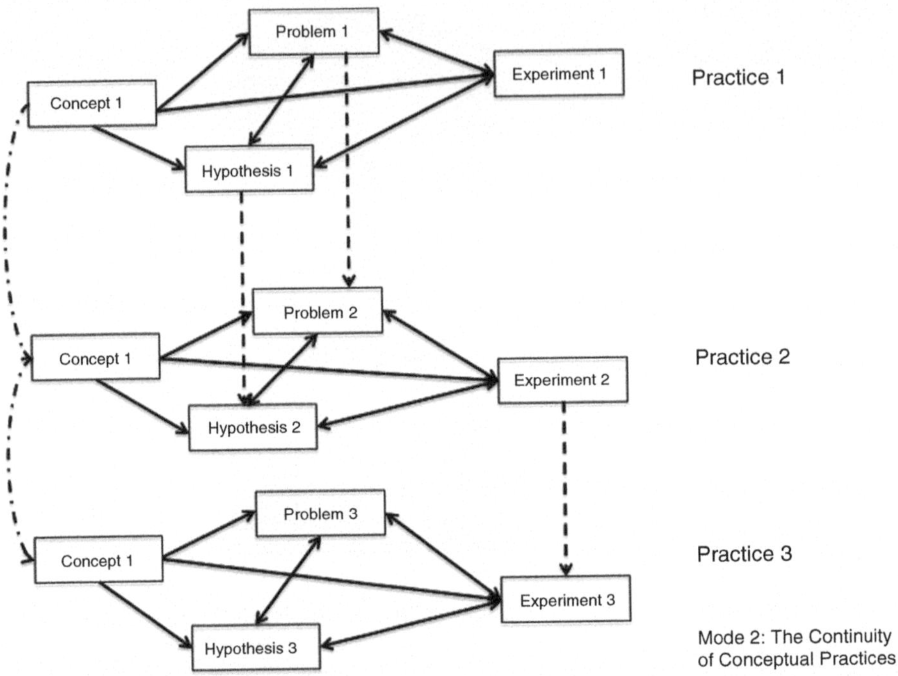

Fig. 8.6 Conceptual continuity mode 2

Thus, based on Mode B of continuity, I propose a new mode of the continuity of scientific practice: A series of scientific practices is continuous if some components are preserved or preserved with modification during some period but it is not necessary for these components to be invariantly preserved throughout history. Given that scientific practice, as I suggested in Chaps. 5 and 6, usually consists of intertwined activities: problem-defining, conceptualisation, hypothesisation, experimentation, and reasoning, the continuity of scientific practices can be illustrated in Fig. 8.6: Some part of hypothesis 1 and problem 1 are preserved with modification in the shift from Practice 1 to Practice 2, while some part of Experiment 2 is preserved with modification from Practice 2 to Practice 3. Although none of any component of activities is invariantly preserved or shared from Practice 1 to Practice 3, it is undeniable that there is some continuity in the shift from scientific practice 1 to scientific practice 3.

As I have argued in Sect. 8.3, conceptual practice should be understood and examined in a broader context. Thus, I propose a new mode of conceptual continuity (Mode 2): *The continuity of the uses of a concept C in history holds if the uses of C contribute to a continuous series of scientific practices.*[9] (Holistic Continuity) As

[9]Note that "the uses of a concept C in history" assumes that the different uses of C are historically linked.

illustrated in Fig. 8.6, Concept 1, despite the substantial change, is still playing a vital role in all of the practice 1, practice 2, and practice 3. It is in this sense that a series of the practices of *C* is continuous because the conceptual change contributes to the continuity of scientific practice in a broader context.

It is worth noting that the crucial difference between Mode 1 and Mode 2 is that Mode 1 requires that some components of a scientific concept are invariantly preserved throughout history, while Mode 2 does not. In Mode 2, there is no part of Concept 1 invariantly preserved from Practice 1 to Practice 3. Nor is any part of other components of activities (e.g. problem-defining, hypothesesation, and experimentation) invariantly preserved from Practice 1 to Practice 3. Nevertheless, all these changes do not impede the continuity from Practice 1 to Practice 3. Nor do they prevent from the continuous practices of Concept 1.

To sum up, I distinguish two modes of the continuity of conceptual practice based on two modes of continuity, and argue that Mode 2 provides a new way to understand the continuity of conceptual practice. I shall illustrate and defend Mode 2 by applying it to the case of the concept of dominance in the next section.

8.5 The Case of the Concept of Dominance Revisited

As shown in Sect. 8.2, the concept of dominance was central to the development of the Mendelian approach to the problem of heredity: Mendel's theory with the concept of dominance was introduced to the study of heredity around 1900. De Vries, Correns, and Bateson all refined the concept and developed the theory accordingly. When Morgan and his colleagues eventually developed the theory into a highly sophisticated form, which was substantially different from the ones formulated by de Vries and Bateson, the concept of dominance was still used to illustrate the simple cases. In the following, I shall show that the uses of the concept of dominance are continuous in early genetics. My argument is as follows.

P1. The concept of dominance was used by Mendel, de Vries, Correns, Bateson, and Morgan in their practices respectively.

P2. Mendel's, de Vries', Correns', Bateson's, and Morgan's work mark a continuous series of exemplary practices in the history of genetics.

P3. A series of the uses of a concept is continuous if the uses of that concept contribute to a continuous series of scientific practices.

C. Therefore, the uses of the concept of dominance were continuous in the history of genetics (from Mendel to Morgan).

Firstly, it is clear that the concept of dominance was used by Mendel, de Vries, Correns, Bateson, and Morgan in their exemplary practices respectively. The concept of dominance was first introduced by Mendel to classify the differing morphological traits by observing the behaviour of hybrids in the successive generations in order to refine the research problem on the development of pea hybrids into a more specific and testable one (MP1 $\rightarrow$ MP2):

What is the law deducible from the changes for each pair of dominant and recessive traits in the successive generations? (Mendel 1866, 7)

Again, based on the concepts of dominance, hybrid form, and recessiveness, Mendel was able to design new experiments (ME1, ME2, ME3) and make a statistical analysis of the results of the experiments in order to study the problem (MP2). Moreover, Mendel's laws (MH1, MH2) as hypotheses were explicitly formulated based on these concepts.

The law of development concerning a pair of differing traits: Of the seeds formed by the hybrids with one pair of differing traits, one half again develop the hybrid form while the other half yield plants that remain constant and receive the dominating and the recessive character in equal shares. (Mendel 1866, 17, 1966, 15)

The law of combination of differing traits: The progeny of hybrids in which several essentially different [dominating and recessive] traits are united represent the terms of a combination series in which the series for each pair of differing traits are combined. And the behaviour of each pair of differing traits in a hybrid association is independent of all other differences in the two parental plants. (Mendel 1866, 22, 1966, 22)

Although Mendel never explicitly attributed "dominant/recessive" to the cell-types as hereditary material, the morphological-cellular correspondence stated in the symbolic formulation (i.e. $\frac{A'}{A'} + \frac{A'}{a'} + \frac{A'}{a'} + \frac{a'}{a'} = A + 2Aa + a$) of his law of composition of fertilising cells (MH3) implicitly suggests a potential usage of "dominant/recessive", which influenced de Vries', Correns' and Bateson's usages. De Vries, on the one hand, following Mendel's dominance$_{\mathrm{Mm}}$, used "dominant/recessive" to refine his central research problem (DP1 → DP2) and re-analyse the data from his hybridisation experiments (DE1, DE2). On the other hand, de Vries proposed a hypothesis (DH6), namely, the law of segregation, in which "dominant/recessive" are attributed to the hereditary characteristics (1900b, 845–46). It is evident that de Vries' symbolic formulation of the law of segregation (i.e. $(D + R)(D + R) = D^2 + 2DR + R^2$) is a revised version of Mendel's (MH3) as a partial solution to the problem.

Correns adopted the terms "dominant" and "recessive" to test Mendel's observation by reformulating Mendel's rule (CH1) and designing the experiments (CE1, CE2). Bateson used "dominant/recessive" to define his research problem (BP1).

What offspring are to expect if two germ-cells of dissimilar constitution unite in fertilisation? (Bateson 1902, 18–19)

Furthermore, Bateson formulated the Mendelian principles of heredity (BH1, BH2) in terms of dominance and recessiveness.

In the simplest case, suppose a gamete from an individual presenting any character in intensity *A* [dominant] unite in fertilisation with another from an individual presenting the same character in intensity *a* [recessive]. For brevity's sake we may call the parent individuals *A* and *a*, and the resulting zygote *Aa*. What will the structure of *Aa* be in regard to the character we are considering? (Bateson 1902, 18–19)

But again—if the case be Mendelian—the gametes borne by *AB* will be either *A*'s or *B*'s, and the cross-bred *AA*'s breeding together will form *AA*'s, *AB*'s and *BB*'s. Moreover, if as in the

> normal Mendelian case, *AB*'s bear on an average equal numbers of *A* gametes and *B* gametes, the numerical ratio of these resulting zygotes to each other will be
>
> 1*AA*: 2*AB*: 1*BB*. (Bateson 1902, 24)
>
> If the zygote be formed by the union of dissimilar gametes, we may meet the phenomenon of (a) dominant and recessive characters; (b) a blend form; (c) a form distinct from either parent, often reversionary. (Bateson 1902, 26)

As I have shown in Chap. 5, Mendel's concept of dominance not only was important to be used in problem-defining, hypothesisation, and experimentation for his work on the development of hybrids, but also enlightened the study of heredity around 1900. The terms "dominant" and "recessive" were adopted by de Vries, Correns, Bateson, and Morgan to define new research problems, propose new hypotheses, and design and conduct the experiments, though their usages were different from Mendel's in some aspects. In particular, the statistical analysis of the dominant and recessive traits was introduced as a useful tool into the study of heredity, especially by de Vries and Bateson.

Secondly, it is undeniable that Mendel's, de Vries', Correns', Bateson's, and Morgan's work mark a continuous process of practice in the history of genetics. The continuity in the chain of exemplary practices from Mendel to Morgan consists of the acceptance and modification of problem-defining, conceptualisation, hypothesisation, and experimentation from the early practice. Again, it should be emphasised that to say that there is a continuity in the series of practice from Mendel to Morgan does not mean that there are certain components that are accepted and shared invariantly in all of the practices from Mendel to Morgan. As illustrated in Fig. 8.7, de Vries' main research problem (DP2) is directly influenced by Mendel's (MP2) and his law of segregation (DH6) is formulated based on Mendel's law of hybrid development (MH1). Bateson's central research problem (BP1) is defined with the help of de Vries' (DP2) and his Mendelian principles (BH1, BH2) are formulated and developed based on de Vries' hypothesis (DH6). Although no component of Mendel's practice is invariantly preserved by his successors', there is a continuity of practices in early genetics. In addition, it should be noted that the different usages of "dominance" by de Vries, Correns, and Bateson do not prevent them from communicating with each other. De Vries' dominance$_{\mathrm{Dc}}$, Correns' dominance$_{\mathrm{Ca}}$, and Bateson's dominance$_{\mathrm{Ba}}$ are clear examples of the attempts to investigate the physical basis of the hereditary bearer by using the concept of dominance. The early Mendelians' revisions of the concept of dominance also reflect a continuous series of the attempts to study the mechanism of heredity.

Therefore, I conclude that the concept of dominance is a highly useful tool, contributing to a continuous series of exemplary practices in early genetics, though its semantic aspect and practical aspect changed in different research contexts. It is in this sense that the uses of the concept of dominance are continuous.

Before finishing this section, I wish to highlight two advantages of my new account of conceptual continuity over the analytic, the cognitive, and the practice-based accounts. First, my account well characterises and accounts for the continuity in a radical conceptual change, which is a serious difficulty for old accounts. Second,

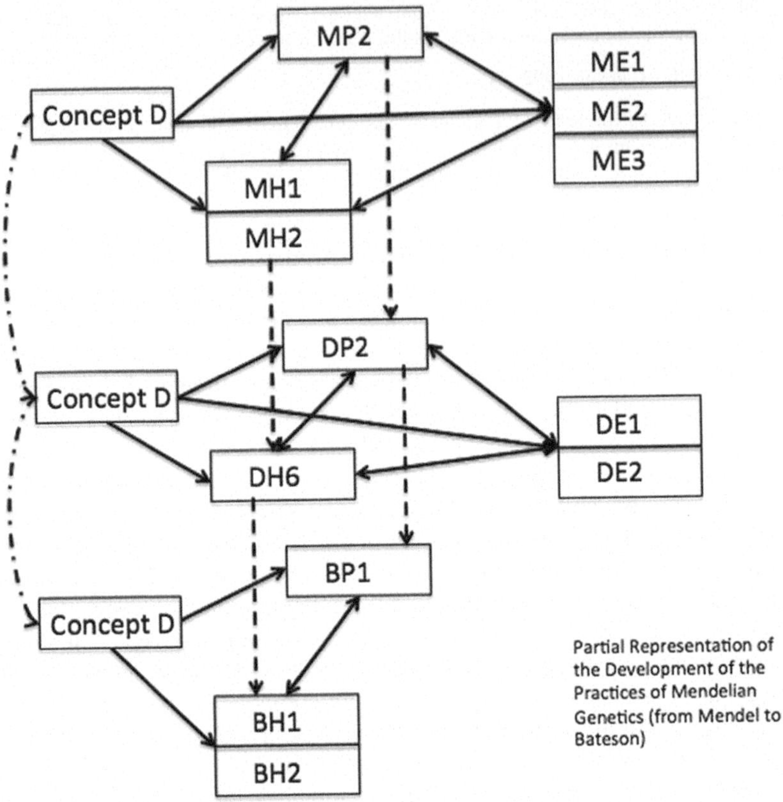

Fig. 8.7 Conceptual continuity in early genetics

my account of conceptual continuity is also helpful to have a fuller understanding of the continuous development of scientific practice in a broader context. As shown above, the holistic analysis of the evolution of the concept of dominance helps us to have a better understanding of the continuous development of genetics in which the concept of dominance plays an important role. This is what is missing from old analyses, which pay too much attention to the continuity of conceptual practice itself.

8.6 Conclusion

To sum up, I argue that an adequate philosophical analysis of conceptual change should be situated in a broader context in which a scientific concept should be understood as a tool to define and solve problems in practice (i.e. by contributing to the activities of problem-defining, hypothesisation, experimentation, and reasoning).

Accordingly, I propose a holistic account of conceptual change, in which the continuity is explainable in a broader context: A series of the uses of a scientific concept is continuous if the uses of that scientific concept contribute to a continuous series of scientific activities in broader context, including problem-defining, hypothesisation, experimentation, and reasoning. With my historical case study of the evolution of the concept of dominance, I show that despite the changes of both the semantic and practical aspects, the continuity of conceptual practices is defensible.[10]

References

Andersen, Hanne, and Nancy J. Nersessian. 2000. Nomic Concepts, Frames, and Conceptual Change. *Philosophy of Science* 67: s234–s241.

Barsalou, L.W. 1992. Frames, Concepts, and Conceptual Fields. In *Frames, Fields, and Contrasts*, ed. A. Lehrer and E.F. Kittay, 21–74. Hillsdale, NJ: Lawrence Erlbaum Associates.

Bateson, William. 1902. *Mendel's Principles of Heredity: A Defence*. Cambridge: Cambridge University Press.

Boon, Mieke. 2012. Scientific Concepts in the Engineering Sciences. In *Scientific Concepts and Investigative Practice*, ed. Uljana Feest and Friedrich Steinle, 219–243. Berlin: de Gruyter.

Brigandt, Ingo. 2010. The Epistemic Goal of a Concept: Accounting for the Rationality of Semantic Change and Variation. *Synthese* 177 (1): 19–40.

———. 2012. The Dynamics of Scientific Concepts: The Relevance of Epistemic Aims and Values. In *Scientific Concepts and Investigative Practice*, ed. Uljana Feest and Friedrich Steinle, 75–104. Berlin: de Gruyter.

Chang, Hasok. 2011. The Persistence of Epistemic Objects Through Scientific Change. *Erkenntnis* 75 (3): 413–429.

Correns, Carl. 1900. G. Mendels Regel über das Verhalten der Nachkommenschaft der Rassenbastarde. *Berichte der Deutschen Botanischen Gesellschaft* 18 (4): 158–168.

———. 1966. G. Mendel's Law Concerning the Behavior of Progeny of Varietal Hybrids. In *The Origin of Genetics: A Mendel Source Book*, ed. Curt Stern and Eva R. Sherwood, translated by Leonie Kellen Piternick, 119–132. San Francisco, CA: W. H. Freeman and Company.

Darden, Lindley. 1991. *Theory Change in Science: Strategies from Mendelian Genetics*. Oxford: Oxford University Press.

Falk, Raphael. 1986. What Is a Gene? *Studies in History and Philosophy of Science* 17 (2): 133–173.

Feest, Uljana. 2010. Concepts as Tools in the Experimental Generation of Knowledge in Cognitive Neuropsychology. *Spontaneous Generations* 4 (1): 173–190.

Gallesio, Count Giorgio. 1816. *Teoria Della Riproduzione Vegetale*. Pisa: Capurro.

Goss, John. 1824. On the Variation in the Colour of Peas, Occasioned by Cross Impregnation. *Transactions of the Horticultural Society of London* 5: 234–236.

Griffiths, Paul E., and Karola Stotz. 2007. Gene. In *The Cambridge Companion to the Philosophy of Biology*, ed. David L. Hull and Michael Ruse, 85–102. Cambridge: Cambridge University Press.

Kindi, Vasso. 2012. Concept as Vessel and Concept as Rule. In *Scientific Concepts and Investigative Practice*, ed. Uljana Feest and Friedrich Steinle, 23–46. Berlin: de Gruyter.

[10]It should be highlighted that the function of the case study of the dominance concept in this chapter is illustrative. The holistic approach is illustrated by this case study rather than generalised from it.

Knight, Thomas Andrew. 1799. An Account of Some Experiments on the Fecundation of Vegetables. *Philosocial Transactions of the Royal Society of London* 89: 195–204.

———. 1824. Some Remarks on the Supposed Influence of the Pollen, in Cross Breeding, upon the Colour of the Seed-Coats of Plants, and the Qualities of Their Fruits. *Transactions of the Horticultural Society of London* 5: 377–380.

Kuhn, Thomas Samuel. 1970. *The Structure of Scientific Revolutions*. 2nd ed. Chicago, IL: The University of Chicago Press.

Mendel, Gregor. 1866. "Versuche über Pflanzenhybriden." *Verhandlungen des Naturforschenden Vereins Brünn* IV (Abhandlungen): 3–47.

———. 1966. Experiments on Plant Hybrids. In *The Origin of Genetics: A Mendel Source Book*, ed. Curt Stern and Eva R. Sherwood, translated by Eva R. Sherwood, 1–48. San Francisco, CA: W. H. Freeman and Company.

Morgan, Thomas Hunt. 1926. *The Theory of the Gene*. New Haven, CT: Yale University Press.

Morgan, Thomas Hunt, Alfred Henry Sturtevant, Hermann Joseph Muller, and Calvin Blackman Bridges. 1915. *The Mechanism of Mendelian Heredity*. New York: Henry Holt and Company.

Nersessian, Nancy J. 1984. *Faraday to Einstein: Constructing Meaning in Scientific Theories*. Dordrecht: Kluwer Academic Publishers.

———. 2008. *Creating Scientific Concepts*. Cambridge, MA: The MIT Press.

Portin, Petter. 1993. The Concept of the Gene: Short History and Present Status. *The Quarterly Review of Biology* 68 (2): 173–223.

Psillos, Stathis. 1999. *Scientific Realism: How Science Tracks Truth*. London: Routledge.

Putnam, Hilary. 1973. Explanation and Reference. In *Conceptual Change*, ed. G. Pearce and P. Maynard, 199–221. Dordrecht: Reidel.

———. 1975. *Mind, Language and Reality*. Cambridge: Cambridge University Press.

Roberts, H.F. 1929. *Plant Hybridization before Mendel*. Princeton, NJ: Princeton University Press.

Sageret, M. 1826. Considérqtions sur la Production des Hybrides, des Variantes et des Variétés en Général, et sur Celles de la Famille des Cucurbitacées en Particular. *Annales des Sciences Naturelles* 8: 294–314.

Sankey, Howard. 1994. *The Incommensurability Thesis*. Aldershot: Avebury.

Seton, A. 1824. Note by the Secretary. *Transactions of the Horticultural Society of London* 5: 236–237.

Stubbe, Hans. 1972. *History of Genetics from Prehistoric Times to the Rediscovery of Mendel's Laws*. Cambridge, MA: The MIT Press.

Thagard, Paul. 1990. Concepts and Conceptual Change. *Synthese* 82 (2): 255–274.

———. 1992. *Conceptual Revolutions*. Princeton, NJ: Princeton University Press.

Toulmin, Stephen Edelston. 1972. *Human Understanding*. Princeton, NJ: Princeton University Press.

von Tschermak, Erich. 1900a. Über Künstliche Kreuzung bei Pisum Sativum. *Berichte der Deutschen Botanischen Gesellschaft* 18 (6): 232–239.

———. 1900b. Über Künstliche Kreuzung bei Pisum Sativum. *Zeitschrift für das Landwirtschaftliche Versuchswesen in Oesterreich* 3: 465–555.

Votsis, Ioannis, and Gerhard Schurz. 2012. A Frame-Theoretic Analysis of Two Rival Conceptions of Heat. *Studies in History and Philosophy of Science* 43 (1): 105–114.

de Vries, Hugo. 1889. *Intracellulare Pangenesis*. Jean: Gustav Fischer.

———. 1900a. Das Spaltungsgesetz der Bastarde (Vorlaufige Mittheilung). *Berichte der Deutschen Botanischen Gesellschaft* 18 (3): 83–90.

———. 1900b. Sur la Loi de Disjonction des Hybrides. *Comptes Rendus de l'Academie des Sciences (Paris)* 130: 845–847.

———. 1900c. Sur les Unités des Caractères Spécifiques et Leur Application à l'étude des Hybrides. *Revue Générate de Botanique* 12: 257–271.

———. 1950. Concerning the Law of Segregation of Hybrids. Translated by Aloha Hannah. *Genetics* 35 (5 (2)): 30–32.

———. 1966. The Law of Segregation of Hybrids. In *The Origin of Genetics: A Mendel Source Book*, ed. Curt Stern and Eva R. Sherwood, translated by Evelyn Stern, 107–117. San Francisco, CA: W. H. Freeman and Company.

Waters, C. Kenneth. 2014. Shifting Attention from Theory to Practice in Philosophy of Biology. In *New Directions in the Philosophy of Science*, ed. Maria Carla Galavotti, Dennis Dieks, Wenceslao J. Gonzalez, Stephan Hartmann, Thomas Uebel, and Marcel Weber, 121–139. Dordrecht: Springer.

Weismann, August. 1892. *Das Keimplasma: Eine Theorie Der Vererbung*. Jena: Gustav Fischer.

Chapter 9
The Gap Problem in Hypothetico-Deductivism

Abstract Facing many famous objections, the hypothetico-deductive (H-D) model is no longer the mainstream account of evidence in the philosophy of science. Nevertheless, in the history (and even contemporary practice) of science, the H-D model has been being widely used. So there is a gap problem between the philosophical analysis of evidence and good actual scientific practice. Achinstein (PSA: Proc Bienn Meet Philos Sci Assoc 67:S180–S192, 2000; Evidence. In: Psillos S, Curd M (eds) The Routledge companion to philosophy of science, Routledge, London, pp 337–348, 2008) attempts to resolve the gap problem by introducing a strong and empirical account of evidence. However, the Achinsteinian solution is still not fully satisfactory. In the chapter, I argue for a practice-based solution to the gap problem. I begin with the case of Mendel's evidence to illustrate the gap problem and show why the Achinsteinian solution fails. Then, I propose an account of a good H-D evidential practice. I further argue that the gap problem can be well resolved by my account.

Keywords H-D confirmation · Evidence · Evidential practice · The gap problem · Mendel

9.1 Mendel's Evidence and the Gap Problem

In his famous paper *Versuche über Pflanzen-Hybriden* (1866), Gregor Mendel proposed a hypothesis to explain the numerical interrelationships between the pairs of differing morphological traits (e.g. round/wrinkle, yellow/green) in the hybrids and their progeny:

MH3. Pea hybrids form germinal and pollen cells that in their composition correspond in equal numbers to all the constant forms resulting from the combination of traits through fertilization. (The law of composition of hybrid fertilising cells).

In order to test MH3 experimentally, Mendel made some testable predictions by deducing from MH3 along with the following assumptions:

Y. Shan, *Doing Integrated History and Philosophy of Science: A Case Study of the Origin of Genetics*, Boston Studies in the Philosophy and History of Science 320,
https://doi.org/10.1007/978-3-030-50617-9_9

P_1. The constant traits can be produced when germinal and pollen cells are alike.

P_2. There are many kinds of germinal cells in the ovaries of hybrids; and there are many kinds of pollen cells in the anthers of hybrids.

P_3. The kinds of the germinal and pollen cells correspond in their internal make-up to the individual forms.

P_4. The seed plants for the experiments all have round shape (denoted as *A*) and yellow albumen (*B*), while the pollen plants all have wrinkled shape (*a*) and green albumen (*b*).

P_5. Four experiments (MEs) are designed as follows:

(1) The hybrid seeds (i.e. with *AB*, *Ab*, *aB*, *ab* traits) are fertilised with the pollen from the plants with *AB* traits.
(2) The hybrid seeds (i.e. with *AB*, *Ab*, *aB*, *ab* traits) are fertilised with the pollen from the plants with *ab* traits.
(3) The seeds from the plants with *AB* traits are fertilised with the hybrid pollen (i.e. with *AB*, *Ab*, *aB*, *ab* traits).
(4) The seeds from the plants with *AB* traits are fertilised with the hybrid pollen (i.e. with *AB*, *Ab*, *aB*, *ab* traits).

It is inferable that during the fertilisation, there are corresponding kinds of germinal and pollen cells that develop and combine.

(1) Germinal cell-types $A'B'$, $A'b'$, $a'B'$, $a'b'$ with pollen cell-types $A'B'$
(2) Germinal cell-types $A'B'$, $A'b'$, $a'B'$, $a'b'$ with pollen cell-types $a'b'$
(3) Germinal cell-types $A'B'$ with pollen cell-types $A'B'$, $A'b'$, $a'B'$, $a'b'$
(4) Germinal cell-types $a'b'$ with pollen cell-types $A'B'$, $A'b'$, $a'B'$, $a'b'$

Mendel further deductively inferred that in the each of the four experiments, the following combinations of the cell-types would be expected and they are in equal ratio to each other.

(1) $\frac{A'B'}{A'B'} : \frac{A'B'}{a'b'} : \frac{A'B'}{A'b'} : \frac{A'B'}{a'B'} = 1: 1: 1: 1$
(2) $\frac{A'B'}{a'b'} : \frac{a'b'}{a'b'} : \frac{A'b'}{a'b'} : \frac{a'B'}{a'b'} = 1: 1: 1: 1.$
(3) $\frac{A'B'}{A'B'} : \frac{A'B'}{a'b'} : \frac{A'B'}{A'b'} : \frac{A'B'}{a'B'} = 1: 1: 1: 1$
(4) $\frac{A'B'}{a'b'} : \frac{a'b'}{a'b'} : \frac{A'b'}{a'b'} : \frac{a'B'}{a'B'} = 1: 1: 1: 1$

Correspondingly, given the MH3 and the P_1, the following traits would be expected to stand in equal ratio to each other in each experiment.

(1) AB: $AabB$: ABb: $aAB = 1: 1: 1: 1$
(2) $AaBb$: ab: Aab: $aBb = 1: 1: 1: 1$
(3) AB: $AaBb$: ABb: $AaB = 1: 1: 1: 1$
(4) $AaBb$: ab: Aab: $aBb = 1: 1: 1: 1$

Moreover, in the experiments (1) and (3), since the dominating traits A and B appeared in every combination, all seeds should be imprinted their characteristics. In other words, it was expected that all seeds obtained from experiments (1) and

Table 9.1 Mendel's experiments

Experiment (1)	Experiment (2)	Experiment (3)	Experiment (4)	Morphological traits
98	31	94	24	Round yellow seeds
0	26	0	25	Round green seeds
0	27	0	22	Wrinkled yellow seeds
0	26	0	27	Wrinkled green seeds

(3) were round and yellow. On the other hand, in the experiments (2) and (4), it was expected that there were four kinds of seeds obtained: round yellow, round green, wrinkled yellow, and wrinkled green.

By undertaking these four experiments (MEs), Mendel obtained the favourable results (see Table 9.1). All seeds obtained in the experiments (1) and (3) were exclusively round and yellow. On the other hand, the amount of the round yellow, round green, wrinkled yellow, and wrinkled green seeds are observed approximately in equal number from the experiments (2) and (4).

Mendel, therefore, concluded that "experimentation justifies" MH3. However, it seems that Mendel was a bit too optimistic. It is really doubtful if Mendel had a good piece of evidence for MH3. From a practical point of view, Mendel's justification is a good case of the hypothetico-deductive model of evidence (H-D evidence), which is typically in the form that *e H-D confirms h relative to k if and only if* $h \wedge k \vDash e$ and $k \nvDash e$.[1] We have seen that the approximated results of MEs H-D confirms MH3 relative to P_1, P_2, P_3, P_4, and P_5, as $MH3 \wedge P_1 \wedge P_2 \wedge P_3 \wedge P_4 \wedge P_5 \vDash$ the approximated results of MEs, and $P_1 \wedge P_2 \wedge P_3 \wedge P_4 \wedge P_5 \nvDash$ the approximated results of MEs. However, it still does not justify that Mendel had good evidence for MH3. The legitimacy of H-D evidence has been seriously troubled by a number of objections, including the problem of irrelevant conjunctions (Hempel 1945; Glymour 1980), the problem of underdetermination (Quine 1951; Duhem 1954; Earman and Salmon 1992), and the problem of statistical hypotheses (Earman and Salmon 1992).[2] For some philosophers (Glymour 1980; Park 2004), given these objections, H-D evidence is hopeless. As Gregor Betz (2013, 991) observes, "[H-D evidence] is nowadays largely dismissed as indefensible".

It is evident that Mendel's justification of MH3 is particularly vulnerable to the problems of underdetermination. The basic idea of underdermination is that from a logical point of view, the same piece of evidence is inferable from an indefinite number of equivalent hypotheses, so even if $h \wedge k \vDash e$ and $k \nvDash e$, it is dubious that e confirms h. The results from MEs can be well derived from de Vries' DH4, DH5, DE1, and DE2. Therefore, given the philosophical criticisms on the H-D theory,

[1]It seems to some that Mendel's evidence is not strictly entailed by MH3 and other assumptions. For example, the 31: 26: 27: 26 ratio was not derived logically. However, in the practice of early genetics, it was acceptable that the 31: 26: 27: 26 ratio could be approximated to be the 1: 1: 1: 1 ratio. And it is not unusual in scientific practice that results of experiments are approximated in order to confirm the hypothesis under test.

[2]I will address these problems in Sect. 9.4.

especially the problem of underdetermination, it is not justified to claim that Mendel had good evidence for MH3.

On the other hand, from many early geneticists' points of view, Mendel's justification was valid and well accepted. Both Hugo de Vries (1900a, b, c) and Carl Correns (1900) were convinced by Mendel's confirmation of MH3, though they revised MH3. What is more, William Bateson (1902, 1909) and T. H. Morgan et al. (1915), Morgan (1926) further developed the theory of heredity on the basis of MH3, which was the foundation of the Law of Segregation in classic genetics.

So, there is a gap between the philosophical analysis of H-D evidence and good actual scientific practice.[3] The philosophical objections to H-D evidence do not impede the good use of H-D evidence in scientific practice.[4] As Peter Achinstein (2000, S182) indicates, "philosophical theories of evidence are ignored by scientists." In response, there have been many attempts to resolve the gap problem in H-D evidence by focusing on making formal modifications (e.g. Gemes 2005; Sprenger 2011; Betz 2013). This chapter proposes a practice-based solution to the gap problem in H-D evidence. In Sect. 9.2, I diagnose the problem of the gap problem and critically review the Achinsteinian solution to it. In Sect. 9.3, I propose an account of H-D evidential practice. In Sect. 9.4, I show how the gap problem in H-D evidence can be resolved by my account of H-D evidential practice. In Sect. 9.5, I discuss a potential objection to my account.

9.2 The Diagnosis of the Gap Problem and the Achinsteinian Solution

From a logical point of view, there are two approaches to the gap problem. One is to bite the bullet by dismissing the legitimacy of scientists' use of H-D evidence. The objections to H-D evidence are so serious and decisive that H-D evidence is hopelessly flawed. Scientists have been simply wrong about what they think of H-D evidence. In Mendel's case, it thus can be argued the results of MEs do not provide any good evidential support for MH3. The other is to revisit the philosophical analysis of H-D evidence. It can be argued that the current philosophical formulation of H-D evidence is not perfect so that it fails to characterise the nature of H-D evidence correctly in order to defend its legitimacy effectively. It can also be argued that the objections to H-D evidence are unfair. It seems to me that the first approach is not very promising, given that so many scientific practices conform to H-D evidence in history. Thus, in this chapter, I shall take the second approach.

[3]It should be noted that not all the uses of H-D evidence in practice count as good ones. What is in tension here is that in actual practice there are some good use of H-D evidence while its legitimacy is dismissed by the philosophical analysis.

[4]Other famous examples of the good use of H-D evidence in scientific practice include Young's double-slit experiment and Einstein's prediction of the perihelion of Mercury.

Achinstein pioneers the second approach to the gap problem. His diagnosis of the gap problem (2000, 2008) is that there are two reasons for why scientists ignore the philosophical theories of evidence.[5]

(a) The philosophical theories of evidence propose a too weak notion of evidence.
(b) The evidential claim is assumed to be *a priori*. Whether *e*, if true, is evidence that *h* is a matter to be determined completely by *a priori* calculation.

For Achinstein, in the case of H-D evidence, even if both $h \wedge k \vDash e$ and $k \nvDash e$ are true, it is insufficient to justify that *e* is a piece of evidence for *h*. Such an account of H-D evidence is too weak in the sense that it "allows the same fact to be evidence for a range of conflicting theories" (Achinstein 2000, S183). For example, the approximated results of ME are not only entailed by $MH3 \wedge P_1 \wedge P_2 \wedge P_3 \wedge P_4 \wedge P_5$, but also by de Vries' law of segregation, Correns' Mendelian Rule, and Morgan's Mendelian principles with corresponding background information respectively. Thus, Achinstein argues that an adequate account of evidence should be stronger in order to well explain in what sense *e* provides a good reason to believe *h*.

In addition, Achinstein (1995, 2000) notes that philosophers typically assume that whether *e* confirms *h* or not is completely determined by an *a priori* calculation (e.g. to check whether $h \wedge k \vDash e$ and $k \nvDash e$, given *e* is true). However, he argues that whether *e* is a piece of evidence of *h* cannot be determined purely *a priori*. In addition to the truth of *e*, there is additional empirical information that can affect the truth of the evidential claim that *e* confirms *h*. This can be illustrated by one of Achinstein's favourite examples, the archaeological evidence for the earliest campfires. It was widely accepted that the burned animal bones in the same layer as stone tools and sediment that looks like wood ash (E) is evidence that the first culinary campfires were built by Peking Man in caves in China between 200,000 and 500,000 years ago (T). However, a group of scientists challenged this in 1998, because they found that the sediment was not wood ash but fine minerals and clay deposited by water. Achinstein emphasises that the recent scientists did not dispute the claim that burned animal bones and sediment that looks like wood ash were found in caves in China. That said, even if E is true, $T \wedge K \vDash E$ and $K \nvDash E$,[6] it still does not justify that E is a piece evidence that T. In addition to the truth of E, some additional empirical information (e.g. the correct interpretation of E[7]) is required in order to justify the evidential claim that E confirms T. Therefore, it is in this sense that the evidential claim that *e* confirms *h* relative to *k* should be empirical, since whether *e* confirms *h* is to be determined by empirical investigation of *e*, *h*, and *k*.

---

[5]Note that Achinstein's diagnosis of the gap problem is not just about the gap problem of the H-D theory of evidence. Rather it is applicable to the philosophical theories of evidence in general.

[6]K in this context is minimally formulated as a hypothesis that if Peking Man set the earliest campfires 200,000–500,000 years ago in caves in China, then the burned animal bones remain in the same layer as stone tools and sediment that looks like wood ash if there is any.

[7]E was interpreted as that sediment that looks like wood ash is wood ash, while recent scientists interpreted E as that sediment that looks like wood ash is fine minerals and clay deposited by water.

Accordingly, Achinstein (2008) suggests that an adequate philosophical account of evidence should satisfy two assumptions:

(A) e is evidence that h, given k, only if, given k, e provides a good reason to believe h.
(B) For at least some e, k, and h, whether e, if true, is evidence that h, given k, is an empirical issue. It can be determined at least in part, by empirical investigation of facts pertaining to e, h, and k.

Thus, an Achinsteinian account of H-D evidence can be proposed accordingly as follows[8]:

e HD-confirms h relative to k if and only if

(i) $h \wedge k \vDash e$ and $k \nvDash e$
(ii) $P(E(h, e)/e \wedge k) > 1/2$, where $E(h,e)$ means that there is an explanatory connection between h and e
(iii) e and k are true

The Achinsteinian account adds two additional conditions to the traditional account. For Achinstein, condition (ii) satisfies assumption (A) as follows. To say that e provides a good reason to believe h given k is to say that $P(h/e \wedge k)$ is sufficiently high and e cannot be a good reason to believe $\neg h$. Thus, it follows that e provides a good reason to believe h only if $P(h/e \wedge k) > 1/2$. This can be shown by supposing that e provides a good reason to believe h given k if the probability that there is an explanatory connection between h and e is greater than 1/2 (i.e. $P(E(h, e)/e \wedge k) > 1/2$). Then, e is evidence that h, given k, only if $P(E(h, e)/e \wedge k) > 1/2$. As $P(E(h, e)/e \wedge k) = P(h/e \wedge k) \times P(E(h, e)/h \wedge e \wedge k)$, it can be shown that e is evidence that h, given the background k, only if $P(h/e \wedge k) > 1/2$. In addition, condition (iii) satisfies assumption (B) in the sense that the truths of e and k, have to be determined by empirical investigation. Thus, the Achinsteinian account of H-D evidence seems to solve the gap problem by proposing a stronger and empirical account of H-D evidence.

I am sympathetic to Achinstein's diagnosis of the gap problem to some extent. The traditional philosophical account of H-D evidence is too weak and the relation of the evidence and the hypotheses should be studied empirically rather than purely *a priori*. If $h \wedge k \vDash e$ and $k \nvDash e$ entails that e H-D confirms h, then, as Achinstein (2000, S186) points out, "it is too easy to get evidence." Thus, some additional conditions have to be fulfilled. Moreover, regarding H-D evidence, philosophers have paid too much attention to logic. But evidential reasoning is more than logical reasoning. Whether e H-D confirms h is not merely an issue of logic. It also involves how e is reliably justified and properly interpreted, which is only viable by empirical

[8] Achinstein does not propose a new account of H-D evidence. Rather he is talking of evidence in general. The Achinsteinian account of H-D evidence is formulated on the basis of his new definition of evidence (2008, 345–47).

investigation. But the Achisteinian account of H-D evidence is not fully satisfactory to this end. Firstly, that *k* is true is not a necessary condition for the evidential claim that *e* confirms *h* relative to *k*. As I shall illustrate in Mendel's case below, even if *k* is not true, it does not undermine the legitimacy of the use of H-D evidence. Secondly, condition (iii) is yet to be further articulated. It is promising to suggest that *e* and *k* are empirically true is a necessary condition, but Achinstein is implicit on in what way *e* and *k* can be shown to be true empirically.

Moreover, the Achinsteinian account is not descriptively adequate. Consider Mendel's case. It is obvious to us today that there are no such things called cell-types as hereditary material postulated by Mendel in P_2 and P_3 as background information. Even in the 1900s, none of the "rediscoverers" accepted the truth of P_2 or P_3. It was never restated without modifications in their papers. Correns (1900) reformulated MH3 as the part of his Mendel's Rule in terms of anlagen rather than cell-types, while de Vries' law of segregation (1900a, b, c) was formulated in terms of hereditary characteristics. Therefore, even according to this Achinsteinian H-D account, Mendel did not have good evidence for MH3, since P_2 and P_3 are not true. However, Mendel did in fact have good evidence for MH3! In other words, the Achinsteinian account still fails to well characterise good actual scientific practice like Mendel's confirmation of MH3.[9]

Therefore, I argue that the Achinsteinian account of H-D evidence still fails to well resolve the gap problem, because it does not adequately fit good actual scientific practice.[10] Nevertheless, his diagnosis is right in the sense that we need a stronger account of H-D evidence and the evidential claim cannot be determined by a purely *a priori* calculation. These two points will serve as the starting point of my articulation of H-D evidence.

---

[9]Some may find my objection unfair in the sense that Achinstein does not require or seek such descriptive adequacy because his goal is not to take all cases in which specific scientists have held some *e* to have been evidence for some *h* and justify their contention in retrospect, rather it is to find a reliable way to assess their evidence claim. However, it should be noted that descriptive adequacy that I discuss here is not a historical concept. I do not assume that a good account of H-D has to account for all the uses of H-D evidence in history adequately. The reason that I object Achinsteinian account is that it fails to justify some good use of H-D evidence like Mendel's, rather than that it fails to account for any use of H-D evidence in actual practice. The Achinsteinian account, if correct, should be able to account for all the good uses of H-D evidence in actual practice. In other words, any adequate account of H-D evidence has to well characterise all instances of good use of H-D evidence. But the Achinsteinian account clearly fails, at least in Mendel's case. It is in this sense that the Achinsteinian account is descriptively inadequate.

[10]This is not a full examination of the Achinsteinian solution. As I shall argue, in order to solve the gap problem in hypothetico-deductivism, one has to show two things: the revised account of H-D evidence fits good actual scientific practice; and the revised account is philosophically defensible in the sense that it well resolves the old problems of H-D evidence (e.g. the problem of irrelevant conutions, the problem of underdetermination, and the problem of statistical hypotheses). Thus, a full examination of the Achinsteinian solution should also encompass a discussion on its responses to the old problems. However, as the Achinsteinian solution does not well fit good actual scientific practice, it simply fails to solve the gap problem, whether it can solve the old problems or not. Thus, I do not delve into a more detailed discussion here.

9.3 A New Solution: From Evidence to Evidential Practice

Recently there has been an ongoing practical turn in the philosophy of science. For many philosophers (e.g. Giere 2011; Chang 2014; Soler et al. 2014; Waters 2014), the central issue in contemporary philosophy of science should shift from what scientists find out to how scientists find out, and from scientific (theoretical) knowledge to scientific practice. To this end, I suggest that, when studying the nature of evidence, it is also the time to shift our attention from what a piece of good evidence is to how a good evidential practice is conducted, and what counts as a good evidential practice. Accordingly, in order to resolve the gap problem of H-D evidence, I propose that we should focus on what a good H-D evidential practice is rather than what a piece of good H-D evidence is.

Before delving into the conditions of a good H-D evidential practice, I find it necessary to articulate the desiderata of a good account of evidential practice in general. I suggest that, in addition to Achinstein's two assumptions, a good account of evidential practice has also to be conditioned by the following assumptions:

(C) Pluralist Assumption: There is a plurality of good evidential practices in science. There might be different types of good evidential practice. Some philosophers (e.g. Gemes 2005) have already suggested that there is no universal theory of evidence. In a similar vein, I argue that we should be pluralistic on the nature and the patterns of good evidential practice. For example, palaeontologists' evidential practice by examining fossil records is genuinely different from physicists' evidential practice by conducting experiments. In particular, the latter relies on the reliability and repeatability of experimentation, while the former usually does not. Thus, I argue that it is not wise to assume that there is a single theory which can capture all the patterns of good evidential practice in science.[11]

[11]It seems to some that this assumption entails a pluralistic stance on evidential practice, which is the claim that there is not a single correct account of evidential practice in science, they could be multiple mechanisms at play in different scientific contexts. We do not need to choose between the hypothetico-deductivist, the Hempelian (Hempel 1945), the Achinsteinian (Achinstein 2008), or the Bayesian (e.g. Carnap 1950; Howson and Urbach 1989) theories, because we can find cases in which each of these is in use. I readily embrace the pluralistic stance on evidential practice, and contend that this pluralistic stance echoes both descriptive pluralism (Kellert et al. 2006) and normative epistemic pluralism (Chang 2012) in science. On the one hand, I counter the monist approach to analysing and interpreting evidential practice in science. The plurality of evidential practice in history provides evidence for the pluralist stance. As Julian Reiss (2014) argues, evidential practice is contextual in nature. The contextual nature of evidential practice suggests the plurality of interpreting evidential practice. Thus, there is no need of searching for a single, complete, and comprehensive theory of evidential practice.

On the other hand, I argue that it is not enough to argue that there is a plurality of approaches to interpreting evidential practice. We should encourage a plurality of evidential practice in actual empirical investigation, as it promotes the benefits of toleration and of integration. For example, qualitative and quantitative hypotheses being confirmed by different accounts of evidential practice shows the satisfaction of different aims of scientific practice. The hypotheses confirmed by different

(D) Contextualist Assumption: Whether an evidential practice is good or not should be assessed within in its broader context, that is, a corresponding exemplary practice. An exemplary practice is rooted in a particular historical context. A once-good evidential practice in some exemplary practice can be regarded as a bad one in a different historical context due to the change of background knowledge and technology. For example, some old concepts in an evidential practice in history may have been abandoned for a long time. We should not be whiggish to evaluate the legitimacy of an evidential practice. Thus, in order to determine whether an evidential practice is good, we have to understand and evaluate it in its historical context.[12]

(E) Fallibist Assumption: A good evidential practice alone does not warrant the truth of the hypothesis or sufficiently provides a criterion of theory-choice. In contrast, what a good evidential practice provides is a sort of justification of the reliable correlation between the evidential practice and the evidential claim. Whether the hypothesis under test is true or should be favoured over other alternatives is a different issue.

Based on these assumptions and Achinstein's assumptions (A) and (B), I propose an account of a good H-D evidential practice by identifying three jointly sufficient conditions. An H-D evidential practice p is a good practice if p fulfills the following conditions in order to show e confirms h.

(I) The logical condition: In p, the evidential proposition e is a logical consequence of the hypothesis under test h and the background information k (i.e. $h \wedge k \vDash e$), while the background information k alone does not entail the evidential proposition e (i.e. $k \nvDash e$);

(II) The practical condition: p is a series of repeatable scientific activities (e.g. conceptualisation, experimentation, and reasoning), which reliably and accurately justifies the evidential proposition e.

(III) The contextual condition: In p, the interpretation of the evidential proposition e and background information k are acceptable and all of e, k, and h are relevant within its historical context.

The logical condition captures the distinctive logical feature of the H-D evidential practice. The practical condition highlights the empirical nature of H-D evidential practice. The justification of e depends on a series of repeatable activities, such as designing and undertaking the experiments or observations, and analysing and interpreting the experimental data. For example, if e is experimentally justified,

accounts of evidential practice can also be the basis for co-optation to achieve the aims of different system of knowledge. Thus, cultivating the multiple approaches to evidential practice contributes to the ultimate aim of Hasok Chang's pluralism (2012, 290), namely, "to improve science by cultivating multiple systems of knowledge."

[12]Reiss (2014) develops a different analysis of the contextual nature of evidence.

then the procedure of the experiments should be repeatable.[13] As I have argued, a complete account of evidential practice cannot overlook how *e* is accurately and reliably justified. And the repeatability of the practice shows some kind of accuracy and reliability. In addition, the practical condition also reinforces the logical condition to constitute a stronger account of H-D evidence than what is proposed by the traditional account. More precisely speaking, the repeatability of evidential practice provides a good reason for an agent to accept *h* on the basis of *e*, in addition to follow the logic of the H-D account. In short, the practical condition satisfies assumptions (A) and (B). The contextual condition fulfills the conceptualist and the fallibist assumptions by suggesting that conceptualisation and interpretation of *e* and *k* in the evidential practice have to be consistent with the knowledge in its historical context. To sum up, this account of H-D evidential practice satisfies assumptions (A), (B), (D), and (E) of an empirical and strong account of H-D evidential practice.

My account differs from the Achinsteinian account in three main aspects. Firstly, condition (iii) is modified and improved by my conditions (II) and (III). The truth of *e* is explicated as the reliable and accurate justification of *e* by repeatable empirical investigation. Condition (II) not only highlights the empirical nature of *e*, but also articulates in what way *e* is empirically justified. The truth of *k* is explicated as the acceptance of *k* within a given historical context. It is clear that condition (III) is better than condition (iii) to capture the contextual nature of evidential practice. Secondly, condition (ii) is abandoned by my account. Achinstein (2000, S184) contends that "evidence is related to probability." It is why he tries to reformulate the claim that *e* provides a good reason to believe *h* in such a probabilistic way. I agree with Achinstein on the point that a probabilistic approach to evidence has many virtues and benefits. However, I do not see the necessity of articulating all kinds of evidential practices in a probabilistic way. As I argued in assumption (C), we should be pluralistic on the nature and the patterns of good evidential practice. Moreover, I cannot see why a probabilistic approach is necessary or indispensable to articulate H-D evidential practice. Conditions (I), (II), and (III), I contend, constitute a strong enough account of good H-D evidential practice. Finally, my account articulates how a good H-D evidential practice is conducted, which is missing from the Achinsteinian account.

In the next section, I shall show how my account of good H-D evidential practice can resolve the gap problem.

[13]It should be noted that experimentation is not an indispensable component of evidential practice. For example, in some historical sciences, whether an evidential practice is good or not does not often depend on the repeatability of experimentation. Nevertheless, a good evidential practice in historical sciences still requires the repeatability of empirical investigation in some way. Some newly discovered fossils could be evidence for a given palaeontological theory, though the practice of the discovery itself is not simply repeatable. But in principle, a similar discovery is repeatable. If there are no further similar fossils discovered in some similar circumstance, it is dubious that the fossils provide any strong evidential support.

9.4 The Gap Problem Revisited

In order to solve the gap problem in hypothetico-deductivism, one has to show two things: the revised account fits good actual scientific practice; and the revised account is philosophically defensible in the sense that it well resolves the old problems of H-D evidence. Therefore, I shall show that my account of evidential practice can achieve both goals.

9.4.1 The Defence of Descriptive Adequacy

I contend that good scientific practice in which H-D evidence is well employed can be well illustrated by my account. For example, Mendel's confirmation of MH3 was well corroborated. Firstly, it is beyond dispute that the Mendel's evidential practice for MH3 is a good H-D one. As I have explicated in Sect. 9.1, Mendel's evidence is logically derivable from $MH3 \wedge P_1 \wedge P_2 \wedge P_3 \wedge P_4 \wedge P_5$, while they cannot be entailed by P_1, P_2, P_3, P_4, and P_5. Secondly, Mendel's H-D evidential practice for MH3 in his exemplary practice had been repeatedly undertaken by his successors. Notably, Correns (1900) repeated and confirmed Mendel's H-D evidential practice step by step. De Vries (1900a, b, c) even applied Mendel's evidential practice to H-D confirm the law of segregation in a wider scope. Thirdly, Mendel's conceptualisation of the data, including distinguishing dominant with recessive characters of peas, and interpretations of the evidence on the numerical interrelationships of the different combinations of the dominating and recessive characters were well acceptable within its historical context. For example, both the 31: 26: 27: 26 ratio and the 24: 25: 22: 27 ratio can be reasonably approximated to be the 1: 1: 1: 1 ratio. In addition, though cell-type as a hypothetical entity does not exist from a present-day viewpoint, it was reasonably proposed to explain the mechanism of transmission at the time. And MH3 is relevant in the historical context of Mendel's exemplary practice, as I have shown in Chap. 5. Thus, it is not hasty to conclude that my account of evidential practice well fits good actual scientific practice.

9.4.2 The Defence of Philosophical Adequacy

In addition, I shall show that my account of H-D evidential practice is philosophically defensible in the way that it can well resolve the old problems of H-D evidence. As I mentioned briefly in Sect. 9.1, there are three main objections to H-D evidence: the problem of irrelevant conjunctions, the problem of underdeterminaiton, and the problem of statistical hypotheses. I shall turn to these problems respectively in the following.

9.4.2.1 The Problem of Irrelevant Conjunctions

The problem of irrelevant conjunctions is typically formulated as follows. If h is H-D confirmed by a piece of evidence e (relative to any k), h and x is H-D confirmed by the same e for an arbitrary x that is consistent with h and k. We can easily check the three conditions for H-D evidence: First, by assumption, h, k, and x are consistent. Second, if $h \wedge k \vDash e$ then also $h \wedge k \wedge x \vDash e$ because logical implication is monotonous with regard to the antecedens. Third, k alone does not entail e because we already know that e H-D-confirms h relative to k. Thus, tacking an arbitrary irrelevant conjunct to a confirmed hypothesis preserves the evidential relation. However, it sounds absurd to claim that "light bends and the earth is a disc" is confirmed by Eddington's observation.

In response, I argue that the contextual condition clearly excludes the possibility of any irrelevant conjunct in the hypothesis under test, because all of e, k, and h should be relevant within its historical context. However, some may press on: Is there a reliable and consistent way to distinguish the relevant hypotheses from the irrelevant ones? In contrast to many formal attempts (e.g. Schurz 1991, 1994; Gemes 1993, 2005; Sprenger 2011),[14] I wish to argue for a simpler but non-formal solution.[15] Pierre Duhem (1954, 216) famously argues that "good sense is the judge of hypotheses which ought to be abandoned." In a similar line, I argue that good sense is the judge of hypotheses which are irrelevant. By good sense, I refer to what Duhem (1954, 217) characterises as "reasons which reason [does] not know" and "mind of fitness." Duhem's solution to problem of theory choice has been widely criticised for his vague definition of good sense. However, I do not think that the criticism is also applicable to a Duhemian account of relevance in terms of good sense here. Whether a hypothesis is relevant to the study is more obvious than which hypothesis should be favoured over the alternatives in theory choice. Most scientists may have good sense to tell whether there is any irrelevant conjunct in the hypothesis under test. It is easy by good sense for a scientist to tell whether a consistent conjunct like "earth is a disc" is relevant or not to his study of heredity. Thus, I argue that the

[14]Most of the formal definitions of relevance are controversial. For the critical examination, see Ken Gemes (1998), Suck-Jung Park (2004), and Luca Moretti (2006).

[15]It should be noted that here I am not trying to show that my non-formal solution is better than all the formal solutions. Rather what I argue is that the problem of irrelevant conjunctions can be resolved by my account of H-D evidential practice.

contextual condition with the Duhemian account of relevance well resolves the problem of irrelevant conjunctions.[16, 17]

9.4.2.2 The Problem of Underdetermination

The problem of underdetermination[18] goes like: Whenever an evidence *e* H-D confirms a given hypothesis h_0, it also confirms infinitely many other hypotheses $h_1, h_2, h_3, \ldots h_n$ that are incompatible with h_0, from a logical point of view. In that case, how can we maintain that the test confirms h_0 in preference to an infinite number of other possible hypotheses $h_1, h_2, h_3, \ldots h_n$? In other words, is h_0 really H-D confirmed by *e*?

As I have argued, *e* should not be regarded as a piece of evidence if it is merely shown to be a logical consequence of the hypothesis under test and the background information. The practical condition and the contextual condition have also to be fulfilled. Thus, even if $h_0 \wedge k \vDash e$ and $k \nvDash e$, $h_1 \wedge k \vDash e$ and $k \nvDash e$, $h_2 \wedge k \vDash e$ and $k \nvDash e, \ldots$ $h_n \wedge k \vDash e$ and $k \nvDash e$, it does not imply all of $h_0, h_1, h_2, \ldots h_n$ are equally confirmed by *e*. It is necessary to examine if the conceptualisation and interpretation of *e* and *k* are acceptable and each of $h_0, h_1, h_2, \ldots h_n$ is relevant within its historical/conceptual context. Of course the problem might be pressed further: What if there are multiple hypotheses which all fufill the logical, practical, and contextual conditions? Does this mean that all of these hypotheses are confirmed by *e*? A short answer is yes. I

[16]Some may argue that introducing the Duhemian account of relevance is an ad hoc move to defend the contextual condition, as it does not provide any independent justification. In response, I argue that the problem of irrelevant conjunctions is a purely logical problem. From a practical point of view, this problem seldom bothers working scientists in history. Scientists are well aware of what hypotheses are relevantly under test. Thus, to some extent the problem of irrelevant conjunctions is "artificial." I do not think that it is a serious flaw that an "artificial" problem is resolved in an ad hoc way.

[17]A related problem, the problem of irrelevant disjunctions, can be solved in a similar way. The problem is typically formulated as follows. If *h* is H-D confirmed by a piece of evidence *e* (relative to any *k*), *h* is H-D confirmed by *e* and *e'* for any arbitrary *e'*. If so, then we may end up with some unacceptable evidential claims like "'the moon is made of cheese' is confirmed by 'the moon is made of cheese or London is in England'." Such a problem is solved in a similar way to that in the problem of irrelevant disjunctions. The contextual condition clearly excludes the possibility of any irrelevant disjunct in the evidence.

[18]The problem of underdetermination can be articulated in two different ways, though the basic idea originates from Duhem (1954) and W. V. Quine (1951). One is also called the problem of evidential holism, which states that since no single hypothesis can be tested in isolation, it seems that no single hypothesis ever entails an empirical prediction. Thus, it cannot be H-D confirmed. This objection can be easily countered by arguing that there is a distinction between hypotheses in use and hypotheses under test, as Joe Morrison (2010) suggests: We can test the hypotheses under test against the hypotheses in use and test the hypotheses in use against others in order to ensure that the hypothesis under test is H-D confirmed. What I focus on here is the other version of the problem of underdetermination, which is sometimes called the problem of alternative hypotheses (Earman and Salmon 1992).

emphasise that the problem of evidence should not be confused with the problem of theory choice. The problem of evidence is literally about in what sense a piece of evidence *e* confirms a hypothesis *h*. However, how to compare and evaluate the competing hypotheses is a different issue. What my account of H-D evidential practice can offer is a way to tell whether *e* confirms *h* rather than whether h_1 is better than h_n. If there are multiple hypotheses which all fufill the logical, practical, and contextual conditions, then it can be concluded that all these hypotheses are H-D confirmed. Nevertheless, it is still an open question on which hypothesis should be chosen as the best one to be accepted. And this question does not pose a serious challenge to my account of H-D evidential practice.

9.4.2.3 The Problem of Statistical Hypotheses

The basic idea of the problem of statistical hypotheses is that H-D evidence fails to deal with the statistical hypotheses. H-D evidence relies on a deductive system of logic, while the statistical hypotheses can only be confirmed inductively. Thus, the confirmation of a statistical hypothesis is not accountable by H-D evidence.

In response, I argue that this problem is based on the assumption that H-D evidence is the only correct and complete account of evidence. However, as I have suggested, this assumption should be abandoned. I argue that a plurality of approaches to evidential practice should be encouraged. That said, my account of H-D evidential practice does not have to provide an articulation of how a statistical hypothesis is confirmed. Thus, the problem of statistical hypotheses should not be a "fundamental problem for the H-D method."

To sum up, I argue that my account of good H-D evidential practice well resolves the gap problem, since it well accounts for the cases of good H-D evidential practice and is defensible against the old problems of H-D evidence.

9.5 Normativity and Contextualism in H-D Evidential Practice

It seems to some that showing the descriptive adequacy and philosophical adequacy of my practice-based account of evidential practice is still not sufficient to solve the gap problem. Solving the gap problem also means creating a rigorous tool that scientists can appeal to and find useful in actual practice where they debate whether some data is or is not evidence for some hypothesis. My practice-based account, however, seems to save hypothetico-deductivism by just adding conditions that allow it to be descriptively adequate. In other words, it seems to merely provide an adequate description of good H-D evidential practice. It is too vague and weak to provide a rigorous way to judge good H-D evidential practices. It appears unclear how one can determine when an evidential practice meets or fails to meet conditions

(I), (II), and (III). Moreover, the malleability of my account appears to allow it to justify any H-D evidential practice. Thus, some may ask: Is there any H-D evidential practice in the history of science that could not fall under this account? A dangerous implication is that my account loses an assessment power of evidence and thus surrenders the job of philosopher of science in assessing evidential claims and creates an anything goes account that allows scientists to have evidence whenever they say so simply because they say so.

In response, I argue that my practice-based account is not as weak and flexible as suggested. It does provide a useful framework to evaluate H-D evidential practices. Not only does it well appeal to good H-D evidential practices, but also it accounts for in what ways some H-D evidential practices are bad or illegitimate. In the following, I shall show in a historical case how my account is helpful to assess and account for a bad evidential practice.

The measurement of a non-Doppler redshift in the spectral lines of the white dwarf Sirius B, reported in 1925 by Walter S. Adams, was widely regarded as a strong observational confirmation of the general theory of relativity. More precisely speaking, the gravitational redshift was a clear case of H-D evidence. Arthur Eddington's theoretical prediction of the relativity shift, derived from the theory, was 20 km/s, while Adams' observational result was 21 km/s. However, ironically the original observational measurement of the Sirius B redshift and the theoretical prediction that prompted it were both in error. Therefore, the Sirius B result surely would have been used as strong evidence against General Relativity rather than in support of it.[19]

The redshift is clearly a bad H-D evidential practice, according to my account. It does not satisfy the logical condition, as Eddington's prediction was not a logical consequence of General Relativity and the relevant background information. Nor does it fulfil the practical condition, because Adams' observation was flawed and not repeatable. Therefore, like the Achinsteinian account, my account provides a rigorous tool to evaluate H-D evidential practices. However, unlike the Achinsteinian account, my assessment is context-sensitive. I urge that any evaluation of an H-D evidential practice has to be made in its historical/conceptual context. This is what is missing from the Achinsteinian account as well as the older accounts.

Nevertheless, I have to emphasise that such a contextual analysis of H-D evidential practice does not imply any form of relativism or the "anything goes" account. As shown in the gravitational redshift case, my account does not warrant the legitimacy of all the cases of H-D evidential practice. The contextual condition is quite flexible, while the logical condition and the practical condition are rigorous enough to distinguish good from bad H-D evidential practices. In short, my account does have an assessment power as well as a descriptive power.

---

[19]For a detailed analysis, see J. B. Holberg (2010).

9.6 Conclusion

In this chapter, I have revisited the gap problem in hypothetico-deductivism. I have shown by explicating Mendel's case that there is a gap between good actual scientific practice and the traditional account of H-D evidence: On the one hand, Mendel's H-D evidence for MH3 was widely accepted by his successors around 1900. On the other hand, the account of H-D evidence itself faces serious objections and thus is no longer regarded as a legitimate theory of evidence by philosophers. Thus, it is puzzling why Mendel's evidence was well received in practice while it was not, philosophically speaking, a good piece of evidence. In response, I argue that in order to analyse the nature of evidence, philosophers should shift the attention from looking for an account of evidence to looking for an account of a good evidential practice. On the basis of Achinstein's diagnosis, I have proposed an account of a good H-D evidential practice and have shown that the gap problem can be well resolved by this account.

References

Achinstein, Peter. 1995. Are Empirical Evidence Claims A Priori? *The Bristish Journal for the Philosophy of Science* 46 (4): 447–473.

———. 2000. Why Philosophical Theories of Evidence Are (and Ought to Be) Ignored by Scientists. *PSA: Proceedings of the Biennial Meeting of the Philosophy of Science Association* 67: S180–S192.

———. 2008. Evidence. In *The Routledge Companion to Philosophy of Science*, edited by Stathis Psillos and Martin Curd, 337–348. London and New York: Routledge.

Bateson, William. 1902. *Mendel's Principles of Heredity: A Defence*. Cambridge: Cambridge University Press.

———. 1909. *Mendel's Principles of Heredity*. Cambridge: Cambridge University Press.

Betz, Gregor. 2013. Revamping Hypothetico-Deductivism: A Dialectic Account of Confirmation. *Erkenntnis* 78 (5): 991–1009.

Carnap, Rudolf. 1950. *Logical Foundations of Probability*. Chicago, IL: The University of Chicago Press.

Chang, Hasok. 2012. *Is Water H_2O? Evidence, Realism and Pluralism*. Dordrecht: Springer.

———. 2014. Epistemic Activities and Systems of Practice: Units of Analysis in Philosophy of Science After the Practical Turn. In *Science After the Practice Turn in the Philosophy, History and Social Studies of Science*, ed. Léna Soler, Sjoerd Zwart, Michael Lynch, and Vincent Israel-Jost, 67–79. New York/London: Routledge.

Correns, Carl. 1900. G. Mendels Regel über das Verhalten der Nachkommenschaft der Rassenbastarde. *Berichte der Deutschen Botanischen Gesellschaft* 18 (4): 158–168.

de Vries, Hugo. 1900a. Das Spaltungsgesetz der Bastarde (Vorlaufige Mittheilung). *Berichte der Deutschen Botanischen Gesellschaft* 18 (3): 83–90.

———. 1900b. Sur la Loi de Disjonction des Hybrides. *Comptes Rendus de l'Academie des Sciences (Paris)* 130: 845–847.

———. 1900c. Sur les Unités des Caractères Spécifiques et Leur Application à l'étude des Hybrides. *Revue Générate de Botanique* 12: 257–271.

Duhem, Pierre Maurice Marie. 1954. *The Aim and Structure of Physical Theory*. Translated by Philip P. Wiener. Princeton, NJ: Princeton University Press.

Earman, John, and Wesley C. Salmon. 1992. The Confirmation of Scientific Hypotheses. In *Introduction to the Philosophy of Science*, ed. Merrilee H. Salmon, John Earman, Clark Glymour, James G. Lennox, Peter Machamer, J. E. McGuire, John D. Norton, Wesley C. Salmon, and Kenneth F. Schaffner, 42–103. Indianapolis/Cambridge: Hackett Publishing Company.

Gemes, Ken. 1993. Hypothetico-Deductivism, Content, and the Natural Axiomatization of Theories. *Philosophy of Science* 60 (3): 477–487.

———. 1998. Hypothetico-Deductivism: The Current State of Play; the Criterion of Empirical Significance: Endgame. *Erkenntnis* 49 (1): 1–20.

———. 2005. Hypothetico-Deductivism: Incomplete but Not Hopeless. *Erkenntnis* 63 (1): 139–147.

Giere, Ronald N. 2011. History and Philosophy of Science: Thirty-Five Years Later. In *Integrating History and Philosophy of Science: Problems and Prospects*, ed. Seymour Mauskopf and Tad Schmaltz, 59–65. Dordrecht: Springer.

Glymour, Clark. 1980. Hypothetico-Deductivism is Hopeless. *Philosophy of Science* 47 (2): 322–325.

Hempel, Carl G. 1945. Studies in the Logic of Confirmation (I.). *Mind* 54 (213): 1–26.

Holberg, J.B. 2010. Sirius b and the Measurement of the Gravitational Redshift. *Journal for the History of Astronomy* 41 (1): 41–64.

Howson, Colin, and Peter Urbach. 1989. *Scientific Reasoning: The Bayesian Approach*. La Salle, IL: Open Court.

Kellert, Stephen H., Helen E. Longino, and C. Kenneth Waters. 2006. Introduction: The Pluralist Stance. In *Scientific Pluralism*, ed. Stephen H. Kellert, Helen E. Longino, and C. Kenneth Waters, vii–xxix. Minneapolis/London: The University of Minnesota Press.

Mendel, Gregor. 1866. Versuche über Pflanzenhybriden. *Verhandlungen des Naturforschenden Vereins Brünn* IV (Abhandlungen): 3–47.

Moretti, Luca. 2006. The Tacking by Disjunction Paradox: Bayesianism Versus Hypothetico-Deductivism. *Erkenntnis* 64 (1): 115–138.

Morgan, Thomas Hunt. 1926. *The Theory of the Gene*. New Haven, CT: Yale University Press.

Morgan, Thomas Hunt, Alfred Henry Sturtevant, Hermann Joseph Muller, and Calvin Blackman Bridges. 1915. *The Mechanism of Mendelian Heredity*. New York: Henry Holt and Company.

Morrison, Joe. 2010. Just How Controversial is Evidential Holism? *Synthese* 173 (3): 335–352.

Park, Suck-Jung. 2004. Hypothetico-Deductivism is Still Hopeless. *Erkenntnis* 60 (2): 229–234.

Quine, Willard V.O. 1951. Two Dogmas of Empiricism. *The Philosophical Review* 60 (1): 20–43.

Reiss, Julian. 2014. What's Wrong with Our Theories of Evidence? *Theoria* 29 (2): 283–306.

Schurz, Gerhard. 1991. Relevant Deduction. *Erkenntnis* 35 (1–3): 391–437.

———. 1994. Relevant Deduction and Hypothetico-Deductivism: A Reply to Gemes. *Erkenntnis* 41 (2): 183–188.

Soler, Léna, Sjoerd Zwart, Michael Lynch, and Vincent Israel-Jost. 2014. Introduction. In *Science After the Practice Turn in the Philosophy, History and Social Studies of Science*, ed. Léna Soler, Sjoerd Zwart, Michael Lynch, and Vincent Israel-Jost, 1–43. New York/London: Routledge.

Sprenger, Jan. 2011. Hypothetico-Deductive Confirmation. *Philosophy Compass* 6 (7): 497–508.

Waters, C. Kenneth. 2014. Shifting Attention from Theory to Practice in Philosophy of Biology. In *New Directions in the Philosophy of Science*, ed. Maria Carla Galavotti, Dennis Dieks, Wenceslao J. Gonzalez, Stephan Hartmann, Thomas Uebel, and Marcel Weber, 121–139. Dordrecht: Springer.

Chapter 10
Promisingness in Theory Choice

Abstract This chapter proposes a new criterion of theory choice. I begin with a criticism on a traditional criterion of theory choice. Contra the traditional approach, I argue that theory choice is a situation where scientists are reasoning what theory should be favoured as the most promising theory in the area rather than the one where scientists choose a theory among all the alternatives to be the best theory in the area. Then, I elaborate the concept of promisingness of theories in terms of potential usefulness. Moreover, I compare promisingness with other diachronic criteria, such as Popper's potential progressiveness, Lakatos' predictive novelty, McMullin's P-fertility, Laudan's fertility-promise, Ivani's fruitfulness, and Šešelja et al's pursuit worthiness. Finally, I argue for the promisingness criterion from a normative viewpoint.

Keywords Theory choice · Promisingness · Usefulness · Mendelism · Biometry · Simplicity · External consistency · Fruitfulness

10.1 Theory Choice in Science

Theory choice has been widely discussed based on an implicit assumption that *theory choice in science is a process of choosing the best theory*[1] *among all the available alternatives*. More precisely, this assumption presupposes an unexamined

[1]It has been recognised that theory choice is not a best phrase if theory is construed narrowly, given that choice in scientific change usually involves the commitment to some non-theoretical elements (e.g. values and experimental procedure). Thus, theory here should be understood in a broad sense, referring to a unit of scientific consensus which encompasses both the theoretical and the non-theoretical elements. Thomas Kuhn's "paradigm (as disciplinary matrix)" (1970), Imre Lakatos' "research programme" (1968), Larry Laudan's "research tradition" (1977), and Hasok Chang's "system of practice" (2014) are among the most famous attempts to articulate such a unit of scientific consensus. In order to avoid making any commitment by adopting the phrases like "paradigm choice" and "system choice", I wish to follow the traditional phrase "theory choice", where theory is construed loosely. Accordingly, theories in the chapter refer to scientific

Y. Shan, *Doing Integrated History and Philosophy of Science: A Case Study of the Origin of Genetics*, Boston Studies in the Philosophy and History of Science 320,
https://doi.org/10.1007/978-3-030-50617-9_10

criterion of theory choice: Theory choice is ultimately about how to rank all the available theories in order to choose *the best* one.[2] This goodness criterion is implicitly or explicitly assumed by most, if not all, of the philosophical literature on theory choice without argument (e.g. Kuhn 1977; Baumann 2005; McMullin 2014; Rizza 2014; Morreau 2015). As Alexander Rueger (1996, 264) neatly summarises, "Traditionally, theory choice has been described as the maximization of some value or combination of values that a 'good' theory should possess." Accordingly, when examining theory choice in science, philosophers typically focus on the following questions: What does the best theory mean? What virtues constitute a good theory? What is a best way to evaluate and rank the alternatives in terms of goodness? However, I shall argue that such an approach is misleading, because the goodness criterion is not well established as assumed. This chapter proposes a new criterion of theory choice. In Sect. 10.2, I criticise the goodness criterion from a historical point of view. In Sect. 10.3, I introduce promisingness as an alternative criterion of theory choice. In Sect. 10.4, I compare my notion of promisingness with other diachronic virtues such as Karl Popper's potential progressiveness, Lakatos' predictive novelty, Ernan McMullin's P-fertility, Laudan's fertility-promise, Silvia Ivani's fruitfulness, and Dunja Šešelja, Laszlo Kosolosky, and Christian Straßer's pursuit worthiness. In Sect. 10.5, I defend the promisingness criterion from a normative viewpoint.

10.2 The Choices in the Mendelian-Biometrician Controversy

It has been taken for granted that how to evaluate and rank the competing theories in order to choose the best one is key to theory choice. Thus, the philosophical examinations of theory choice have been overwhelmingly centred on the criteria and rationality of choice by explicating the goodness of a theory: Is there any universal criteria of ranking the alternatives in theory choice (e.g. Kuhn 1977, 1983)? Is it possible to make a rational judgment on theory choice to determine the best option (e.g. Okasha 2011; Bradley 2017)? However, I argue that this approach is to a great extent misleading. History tells us that what scientists ultimately aim at in theory choice is not the best theory at the time.

Early genetics is such a clear case. At the turn of the twentieth century, there was a confrontation between two popular theories of heredity: the Mendelian theory and

consensuses based on some exemplary practices. For example, when talking of the Mendelian theory, I refer it to a scientific consensus based on some exemplary practices in the study of heredity (e.g. Bateson 1902).

[2]Nevertheless, there is a persistent debate on what counts as a best theory in contemporary philosophy of science (e.g. Kuhn 1977; Ivanova and Paternotte 2013; McMullin 2014; Morreau 2014).

the Biometric theory. The Mendelians, led by William Bateson, were optimistic on the future of an approach inspired by Gregor Mendel's work (1866) to the study of heredity, while the Biometricians, led by Karl Pearson and W. F. R. Weldon,[3] developed a statistical approach centred on Francis Galton's law of ancestral heredity (1889). It is worth noting that in the Mendelian-Biometrician controversy, both sides of the contenders also admitted the limitation of their favourite theories. Weldon, for example, was not fully satisfied with the Biometric theory. Though recognising its general applicability to the phenomena of blended inheritance, Weldon was aware that a Biometric theory based on Galton's law of ancestral heredity has a limited application to the phenomena of particulate inheritance and alternative inheritance.

> The work of [Galton and Pearson] has given us an expression for the effects of blended inheritance which seems likely to prove generally applicable, although the constants of the equations which express the relation between divergence from the mean in one generation and that in another may require modification in special cases. Our knowledge of particulate or mosaic inheritance, and of alternative inheritance, is however still rudimentary, and there is so much contradiction between the results obtained by different observers, that the evidence available is difficult to appreciate. (Weldon 1902a, 228)

On the other hand, the limitation of the Mendelian theory was immediately recognised soon after the introduction of Mendel's work to the study of heredity (e.g. Correns 1900; Bateson 1902). Therefore, the Mendelians and the Biometricians agreed on the empirical inadequacy of both theories somehow. Moreover, it seems that the focus of the debates was not whether the Mendelian or the Biometric theory was the better theory of heredity at the time. Bateson (1902), for example, admitted that Mendel's principles were not universally applicable to the problem of heredity, but he firmly believed in the promising future of a Mendelian theory.

> [B]y the application of [the Mendelian] principles we are enabled to reach and deal in a comprehensive manner with phenomena of a fundamental nature, lying at the very root of all conceptions not merely of the physiology of reproduction and heredity, but even of the essential nature of living organisms; and I think that I used no extravagant words when, in introducing Mendel's work to the notice of readers of the Royal Horticultural Society's Journal, I ventured to declare that his experiments are worthy to rank with those which laid the foundation of the Atomic laws of Chemistry. (Bateson 1902, 35)
>
> [The study of the Mendelian system] will provide a most fascinating pursuit, which if followed with assiduous care may at any moment lead to some considerable advance in scientific knowledge. (Bateson 1909, vii)

As Lindley Darden (1977) points out, what Bateson in fact supported is not any particular version of the Mendelian theory of heredity, but rather the promise of Mendelism for further research. In other words, the reason that Bateson favoured Mendelism over Biometry was that Mendelism was more promising than Biometry. Likewise, what the Biometricians, especially Pearson, advocated was a more promising future of the statistical approach to the study of heredity more than the

[3]Here I mean early Weldon (1902–1904). For a detailed analysis of Weldon's view on inheritance, see Chap. 4.

conviction that the Biometric theory was better than the Mendelian theory at the time. They contended that a Galtonian statistical approach was more promising to make the "careful classifications of inheritance for the several grades of relationship, for a great variety of characters, and for many types of life" (Pearson and Lee 1900, 121). Therefore, the Mendelians and the Biometricians did differ in which theory was the most promising.

Some may object this conclusion by arguing that the Mendelians and the Biometricians did dispute over which theory was the better theory of heredity. For example, Bateson (1909, 235) criticised the Biometric theory for its "misuse of statistical method", while Weldon (1902a, 252) dismissed the Mendelian theory for its "neglect of ancestry." However, I argue that these criticisms in fact reinforce my argument. Bateson's objections to the Biometric theory did not only show that the Mendelian theory was better, but also suggested that the Mendelian theory was more promising. Likewise, Weldon's critiques of the Mendelian theory did show his scepticism of both the Mendelian theory and its future. For Bateson (1909, 130), "it is absurd to trace the works of any Law of Ancestral Heredity", while for Weldon (1902b, 55), the Mendelian theory "can only result in harm." In other words, what the Mendelians aimed to show was the Biometric theory was not only worse at the time but also methodologically problematic for future investigation. In contrast, the Biometricians tried to argue that the Mendelian theory was hopeless flawed. Hence, I argue that what the Mendelians and the Biometricians ultimately differed in was which theory was more promising.

Similar conclusions can be drawn from other historical cases. In the Copernican revolution, most of the Copernicans like Galileo and Kepler well recognised the empirical inadequacy of the Copernican theory, but they still advocated it for its promisingness (Kuhn 1957, 209–25). In contemporary physics, those who are working on the string theory are usually fascinated by its promisingness. Thus, it is not hasty to conclude that scientists typically aim at a more promising theory in their theory appraisal. In other words, in contrast to what is traditionally assumed in the goodness criterion, I argue that theory choice is a process of ranking theories in order to choose the most promising one.[4]

[4]That being said, I would like to highlight that my argument should be not understood in the way that scientists did not look for the best theory at the time in theory choice. They did work hard to look for the best available theory sometimes but by doing so, they ultimately aimed at the most promising theory.

10.3 Promisingness as Potential Usefulness

An immediate question now comes: What counts as a more promising theory? I propose that *a theory T_1 is more promising than another T_2 if T_1 is more likely to be more useful than T_2 after a further development of both*. By usefulness, I follow the core idea elaborated in Chap. 6: A theory is useful if and only if it provides some exemplary practices (or, some particular ways of problem-defining and problem-solving) which are repeatable, and provide a reliable framework for further investigation to solve the unsolved problems and to generate more testable research problems across more different fields (or disciplines). Accordingly, *a theory T_1 is more promising than another T_2 if T_1 is more likely to provide some exemplary practices which are repeatable, and provide a reliable framework for further investigation to solve the unsolved problems and to generate more testable research problems across more different areas (or disciplines) than T_2 after a further development of both.*

As I have elaborated in Sect. 6.2, the notion of usefulness encompasses four virtues: repeatability, problem-solving success, problem-defining novelty, and interdisciplinarity. An explanation of the choices in the Mendelian-Biometrician controversy in terms of promisingness fits perfectly well with the exemplar-based analysis of the origin of genetics. Bateson, for example, was confident of the promising future of a Mendelian theory of heredity for its potential usefulness. More precisely speaking, Bateson contended that, with a proper development, a Mendelian theory of heredity was more likely to be more useful to study the problem of heredity. He was even explicit on the point that the Biometric theory would be eventually shown to be useless in the study of heredity.

> Of the so-called investigations of heredity pursued by extensions of Galton's non-analytical method and promoted by Professor Pearson and the English Biometrical school it is now scarcely necessary to speak. That such work may ultimately contribute to the development of statistical theory cannot be denied, but as applied to the problems of heredity the effort has resulted only in the concealment of that order which it was ostensibly undertaken to reveal. A preliminary acquaintance with the natural history of heredity and variation was sufficient to throw doubt on the foundations of these elaborate researches. To those who hereafter may study this episode in the history of biological science it will appear inexplicable that work so unsound in construction should have been respectfully received by the scientific world. With the discovery of segregation it became obvious that methods dispensing with individual analysis of the material are useless. (Bateson 1909, 6–7)

For Bateson, the Mendelian theory was also more likely to be more useful to study the problem of variation.

> [T]hose who as evolutionists or sociologists are striving for wider views of the past or of the future of living things may by the use of Mendelian analysis attain to a new and as yet limitless horizon. (Bateson 1909, 17)
>
> No one who is acquainted with Mendelian method will doubt that by its use practical breeders of animals and plans may benefit. (Bateson 1909, 291)

The Biometricians, on the other hand, contended the potential usefulness of a statistical approach.

[I]f Darwinian evolution be natural selection with heredity, then [Galton's law of ancestral heredity] must prove almost as epoch-making to biologist as the law of gravitation to the astronomer. (Pearson 1898, 412)

Whatever views we hold on selection, inheritance, or fertility, we must ultimately turn to the mathematics of large numbers, to the theory of mass phenomena, to interpret safely our observations. (Weldon et al. 1901, 3)

As many historians (e.g. Provine 1971; Norton 1973; Mayr 1982; Olby 1989) have pointed out, this is one crucial point that the Mendelians differed in from the Biometricians. Bateson argued that a Mendelian theory was more likely to be more useful to account for the nature of variation after a further development, while the Biometricians believed in the greater usefulness of a full-fledged Biometric theory to explain the nature of variation. Thus, for early Mendelians, Bateson's exemplary practice, developed from Mendel's and de Vries', was useful and would be more useful, while Biometricians believed in the potential usefulness of Galton's exemplary practice.

It should be noted that what consists in promisingness is a different issue from what suggests or projects promisingness. As I have proposed, a more promising theory is a theory that is more likely to be more useful than its rival after a further development. In order to evaluate the usefulness of theories, we have to assess repeatability, problem-solving success, problem-defining novelty, and interdisciplinarity. However, in order to assess which theory is more promising, it is not the same as to assess which theory is more useful. More precisely, the judgment that theory T_1 is more useful than another T_2 at a certain time does not imply the judgment that T_1 is more likely to be more useful than T_2 after a further development of both.

I suggest that two virtues typically project the promisingness of a scientific theory: simplicity and external consistency. For scientists, simplicity is usually an indicator of the promisingness of a scientific theory. It should be noted that scientists and philosophers understand simplicity differently. For philosophers (e.g. Forster and Sober 1994; Baker 2003), simplicity is typically regarded as a theoretical virtue, referring to the desire to minimise the number of individual new entities postulated by theories. However, scientists understand simplicity in a broader sense. Not only can theories be simple (in the way suggested by philosophers), ways of experimentation, conceptualisation, and reasoning can also be simple. For example, many early Mendelians were committed to the Mendelian theory due to its simplicity in practice (especially in its simple way of conceptualising morphological traits). As R. C. Punnett (1911) repeatedly highlighted, simplicity was a virtue of the Mendelian theory. It has also been argued that the simplicity of the Copernican theory was a key factor for Kepler and Galileo to accept it as a more promising theory (Kuhn 1977, 323–24). However, it is worth noting that simplicity is a relative notion. What seems to be simple for a scientific community is not necessarily simple for another.

External consistency is another important indicator of promisingness. It refers to the consistency of a theory with a broader scientific framework. In history, scientists tend to favour the theory which is more consistent with his broader scientific commitment over the one which is less consistent. For example, an important reason for early Weldon[5] to strongly oppose the Mendelian theory of heredity was its implication for evolutionary studies: The process of evolution was discontinuous (Provine 1971; Norton 1973; Mayr 1982; Olby 1989). Bateson (1902, 7) explicitly claimed that the phenomena of dominance in F_1 generation were "all examples of discontinuous variation." In contrast, Weldon (1895, 380) contended that "specific modification is at least generally a gradual process." It is also the Biometric standpoint, as summarised by Pearson.

> [T]he main evolution has not taken place by leaps, but by continuous selection of the favourable variation from the distribution of the offspring round the ancestrally fixed type. (Pearson 1908, 39)

The debate on the forms of variation between Weldon and Bateson began in 1894, 7 years earlier than their debate over Mendelism. Bateson (1894, 2) argued that "[T]he forms of living things, taken at a given moment, do nevertheless most certainly form a discontinuous series and not continuous series." Weldon (1894) quickly dismissed Bateson's arguments for discontinuous variations without a substantial and detailed discussion.

> [The arguments] in favour of Mr. Bateson's main contention therefore fail, when applied to any part of the process of evolution of which we can know anything. It remains to consider what experimental evidence is brought forward to prove that variation is in fact "discontinuous" in any living animals... No definition of what exactly is meant by "discontinuous" variation is given.[6] (Weldon 1894, 26)

To some extent, the debate on Mendelism between Weldon and Bateson was an extension of their debate over the forms of variation. As Peter Froggatt and Norman Nevin (1971, 19) point out, "Weldon [was] immersed in mutual problems of inheritance and evolution." This is a clear example of how external consistency plays a role in scientists' judgement on the promisingness of a scientific theory. The difference between Bateson's and Weldon's views also highlights that external consistency is a relative notion. For Bateson, the Mendelian theory is externally consistent with discontinuous variation, while for Weldon, the biometric theory is externally consistent continuous variation.

Thus, I propose that simplicity and external consistency are two virtues that typically project promisingness.[7] That said, it should be noted that I am not trying to argue that these two virtues are sufficient to project prominsingness. In different historical contexts, there might be other virtues that play a role. A judgment of the

---

[5]As I have shown in Sect. 4.3, Weldon's work on inheritance in the first decade of the twentieth century can be divided into two periods: 1901–1904 and 1904–1906. The works in the first period were basically the criticisms of Mendelism.

[6]Note that Weldon accepted that there were the phenomena of discontinuous variation, but he did not think that discontinuous variation played any significant role in the process of evolution. In other words, what Weldon differed from Bateson was the significance of discontinuous variation in the study of evolution.

[7]It should be noted that by arguing that simplicity and external consistency are two virtues that typically project prominsingness, I am not offering an algorithm of theory choice.

promisingness of a theory is historically contingent. There are also some sociological factors which influence scientists' judgments. For example, educational background played a role in the Mendelian-Biometrician controversy. The Biometricians had stronger background in mathematics than the Mendelians in general. Pearson was a world-leading statistician.[8] Weldon also had a background in mathematics. He entered UCL as an undergraduate student in 1876, where he was most interested in and deeply influenced by Olaus Henrici's lectures on mathematics. Moreover, in order to grasp the statistical approach better, Weldon carefully studied statistics and probability in 1889 (Provine 1971, 31). It is clear that Weldon's solid background in mathematics underlay his early advocacy of the Biometric approach to the problem of heredity.[9] In contrast, a lack of adequate mathematical skills was a significant reason for many naturalists to resist the Biometric theory. As Froggatt and Nevin (1971, 28) indicate, many naturalists opposed the Biometricians' methods because they "did not understand them." Bateson seems to be a good example. In contrast to Weldon, Bateson received little training in mathematics. He was struggling to pass Little-go mathematics in his first year at the University of Cambridge. He (1905, 390) even admitted that "[n]eedless to say, my knowledge of mathematics is now nil." Froggatt and Nevin (1971, 31) concludes that this is an important factor of why Bateson never well understood the Biometricians and "instinctively preferred the simpler 'mathematics' of segregation and of observation based on breeding experiments."

10.4 Promisingness, Potential Progressiveness, Potential Fertility, and Fruitfulness

As some may notice, a diachronic criterion of theory choice is not something completely new. Popper (1963) proposes the criterion of relative potential progressiveness (or potential satisfactoriness): Given a pair of rival theories, one should choose the one with more empirically falsifiable content.

> [T]he theory which contains the greater amount of empirical information or content; which is logically stronger; which has the greater explanatory and predictive power; and which can therefore be more severely tested by comparing predicted facts with observations. In short, we prefer an interesting, daring, and highly informative theory to a trivial one. (Popper 1963, 217)

---

[8]Pearson founded the world's first statistics department at UCL in 1911 and has been widely credited as the founder of modern statistics.

[9]In addition, Weldon's professional experience influenced his adoption of the statistical approach. Since 1895, Weldon had closely collaborated with the Marine Biological Association, where the research carried out would well fit with the statistical approach. Moreover, Weldon's close collaboration and friendship with Pearson also reinforced his commitment to the statistical approach.

In a word, Popper suggests that whether a theory is superior to another is not only about its current epistemic virtues, but also about its potential of falsifiability. Imre Lakatos (1978) further develops this falsificationist criterion in terms of predictive novelty.

> That is, I give for criteria of progress and stagnation within a programme and also rules for 'elimination' of whole research programmes. A research programme is said to be progressing as long as its theoretical growth anticipates its empirical growth, that is, as long as it keeps predicting novel facts with some success ... If a research programme progressively explains more than a rival, it 'supersedes' it, and the rival can be eliminated... (Lakatos 1978, 112)

While Lakatos provides a more sophisticated account of the nature of the falsifiable content of a theory, both Popper's and Lakatos' criteria are rooted in a basic idea: Whether a theory is superior to its rival depends on whether it is "riskier" to be falsified.

At first glance, both the falsificationist criteria and my criterion highlight the significance of the diachronic virtues of a theory in theory choice. However, there is a fundamental difference between my concept of promisingness and the falsificantionist concept of potential progressiveness. For Popper and Lakatos, theory choice is fundamentally a process of falsification or elimination. A theory that is judged as "pseudoscientific" (in Popper's word) or "degenerating" (in Lakatos' word) should be ruthlessly abandoned.

However, I still find theory choice a positive activity. Scientists are looking for the most promising theory, while they do not have to reject or abandon those less promising ones. Scientists actively choose to develop more promising theories for further research rather than passively keep the "riskier" theories to be falsified in the future.

Moreover, my criterion better captures the actual practice in the history of science. The reason behind the disagreement between Mendelians and Biometricians was not falsifiability or predictive novelty. For Bateson, the Mendelian theory was superior to the Biometric theory not because of its falsifiability or predictive novelty. At the time, the predictive success of the Mendelian theory was quite limited. And there were not many important novel but untested predictions derived from the Mendelian theory. Rather Bateson believed in a potential of usefulness of the Mendelian theory. On the Biometricians' side, the Mendelian theory was not rejected for its unfalsifiablity. For Pearson and Weldon, the limited applicability of the Mendelian theory suggested a less promising future.

Another influential diachronic criterion of theory choice has been developed in terms of fertility. McMullin (1976), for example, insightfully distinguishes two types of theory appraisal: the epistemic appraisal and the heuristic appraisal. The epistemic appraisal is about the proven fertility of a theory (P-fertility), while the heuristic appraisal is about the as-yet-untested fertility (U-fertility) .[10] McMullin's P-fertility is defined as follows.

> The [P-fertile] theory proves able to make novel predictions that were not part of the set of original explananda. More important, the theory proves to have the imaginative resources, functioning here rather as a metaphor might in literature, to enable anomalies to be overcome and new and powerful extensions to be made. Here it is the long-term proven ability of the theory or research program to generate fruitful additions and modifications that has to be taken into account." (McMullin 1982, 16)

It is clear that there is a crucial difference between P-fertility and promisingness. McMullin's P-fertility is a "proven ability", while promisingness, as I have emphasised, is about potential ability.

Laudan (1981, 152) proposes the concept of potential fertility as the criterion of theory choice. He is explicit on the point that scientists should "seek theories which promise fertility in extending the range of what we can now explain and predict." Nevertheless, my concept of promisingness differs from Laudan's concept of potential fertility in several significant aspects. I do not think that promisingness is identical with a promise of fertility, which, according to Laudan, simply means problem-solving success. I agree with Laudan on the point that problem-solving success really matters, but I do not think that it is the only virtue of a good scientific theory. In other words, a promising theory should be more than a theory with a highly efficient problem-solving power. In addition, unlike Laudan, I do not wish to conflate good decision in theory choice with scientific progress. For Laudan, a good decision in theory choice guarantees scientific progress, because scientific progress is defined in terms of problem-solving power. However, it seems to me that a good decision in theory choice does not necessarily entail scientific progress. It is historically contingent to for a theory to achieve the promise of usefulness.

In addition, I argue that promisingness is a better criterion than potential fertility, from a historical viewpoint. Consider the Mendelian-Biometrician case. Both the Mendelians and the Biometricians were not only optimistic about the potential problem-solving power of their favourite theories, but also about the potential problem-defining novelty and the potential interdisciplinarity. As Bateson (1902, 25) speculated, a full-fledged Mendelian theory would be able to deal with the phenomena of a fundamental nature "not merely of the physiology of reproduction and heredity, but even of the essential nature of living organisms."

Recently, Ivani (2019) develops a new account of fruitfulness, in which fruitfulness is defined as "an ability of research programs to develop." At first glance, there are a few similarities between between Ivani's fruitfulness and my promisingness. First, both highlight the significance of research problems in scientific practice. Ivani argues that "[r]esearch questions have an important role in determining the methods used to conduct the research." This is similar to what I have argued in Sect. 5.4, that is, problems and their solutions are contextually intertwined. Second, both Ivani's

[10]Unfortunately, McMullin (1976, 2014) talks little of U-fertility, because his focus has been P-fertility.

fruitfulness and my promisingness emphasise the significance of the reliability of scientific practice. Ivani argues that "[f]ruitful programs use reliable heuristics", while I propose that a more promising theory is more likely to provide a reliable framework for further investigation.

However, there are substantial differences between Ivani's fruitfulness and my promisingness. First, concerning the significance of research problems in scientific practice, Ivani talks more of problem-choosing (or in her words, "the choice of research questions"), while I refer to problem-defining. I agree with Ivani on the point that research problems are typically starting points of scientific practice and mutually intertwined with the ways of solving problems. However, as I have argued in Chaps. 5 and 6, in many cases, problems are not well pre-defined for scientists to choose. It is a main task for scientists to define a problem actively and creatively. A best theory typically is more likely to provide more useful ways of problem-defining and problem-solving (i.e. more useful exemplary practices) rather than to provide a better way to choose research problems. For example, the Mendelian-Biometrician controversy was not a debate on whether one should choose the Mendelian problems or the Biometric problems as the starting point of the study of heredity. There were few well-defined research problems in the Mendelian framework. Thus, an urgent task for the Mendelians was to look for good ways of defining Mendelian research problems in the study of heredity. In other words, what was at issue is problem-defining rather problem-choosing.

Second, my account of promisingness provides a richer account of reliability than Ivani's fruitfulness. For Ivani, more reliable heuristics suggests more well-designed hypotheses (i.e. more "testable hypotheses that have good chances to be true"). My account of reliability encompasses problem-solving success, problem-defining novelty, and interdisciplinarity. A framework provided by an exemplary practice is reliable if it solves more problems and generates more testable research problems across different fields.

Third, Ivani's work on fruitfulness is still within the context of the good criterion. She regards her account of fruitfulness as "a starting point for a serious discussion of the actual relevance of fruitfulness in the assessment of research programs." In other words, Ivani's fruitfulness aims to help to understand in what sense a theory is the best among its alternatives. However, as I have argued, the promisingness criterion is an alternative to the goodness criterion. We should no longer worry too much which theory is the best and in what sense a theory is the best. Rather we should look for the most promising theory in theory choice.

Another similar concept is "pursuit-worthiness" developed by Šešelja, Kosolosky, and Straßer (2012). A general way to understand "X is worthy of pursuit for Y" is as follows.

> It is rational for Y to pursue X if and only if pursuing X is conducive of the set of goals Z. (Šešelja et al. 2012, 53)

In short, pursuit-worthiness is defined in terms of goal-directed rationality. Such an account implies several fundamental differences between pursuit-worthiness and prominsingness. First, pursuit-worthiness is fundamentally a criterion of rationality, while prominsingness a criterion of theory choice. Surely, pursuit-worthiness as a criterion of rationality can be applied to the problem of theory choice in the following way. If a theory T_1 is rationally pursuit-worthy and another T_2 is not, then one may argue that one should favour T_1 over T_2. However, the problem of rationality and the problem of theory choice are two distinct problems. The former is basically a normative issue, whereas the latter is descriptive as well as normative.

Second, promisingness is a comparative notion in nature, while pursuit-worthiness is not necessarily so. It is pointless to claim that a theory is promising. When talking of the promisingness of a theory T, one should be explicit on its foil: Which theory is T more promising than? However, one can make a non-comparative judgment that a theory is deemed rationally pursuit worthy. For example, Šešelja and Erik Weber (2012) examines the rationality of the theory of contingent drift. Whether the theory of contingent drift is rational or not is not dependent on a comparison between it and other theories.

Third, the judgment that a theory T_1 is rationally worthy of pursuit is not identical with the judgment that T_1 should be favoured over other alternatives. The pursuit worthiness of a theory is not sufficient for one to favour it over other alternatives. For example, if two or more theories are rationally worthy of pursuit, it is underdetermined which theory should be favoured. Even Šešelja, Kosolosky, and Straßer (2012, 57) admit, "It may, for instance, be the case that the evaluator ends up with theories T1, T2 and T3 that are deemed pursuit worthy for Y." In other words, pursuit worthiness is too weak to be a good criterion of theory choice.

Fourth, pursuit-worthiness is defined in terms of goal-directed rationality, while prominsingness in terms of potential usefulness. For Šešelja, Kosolosky, and Straßer, pursuit-worthiness is evaluated in the context with a given set of goals. This account overlooks the significance of investigative (or exploratory) feature of scientific practice. C. Kenneth Waters (2004, 2007, 2014) has argued that scientific practice should be better analysed in terms of "open-ended research." It is problematic to assume that all science is practised with a set of fixed goals. In contrast, my account of promisingness well captures the investigative feature of scientific practice. As I have elaborated in Sect. 10.2, potential usefulness encompass potential problem-defining novelty and interdisciplinarity, which highlights the investigative feature of scientific practice in a nuanced way.

Before finishing this section, I wish to highlight that promisingness cannot be construed as a constituent of goodness. It seems to some that if a theory T_1 is better than another T_2, then T_1 is more promising than T_2. However, I argue that there is a crucial difference between promisingness and goodness. Goodness is an actual property, while promisingness is a modal property. To say that a theory T_1 is better than T_2 is to say that T_1 is now superior to T_2 in general, given all the aspects considered. However, to say a theory T_1 is more promising than T_2 is to say that T_1 is more likely to be "superior" to T_2 in general with further development in the future.

10.5 Argument from Normativity

Another important issue is yet to be examined. Even if I have convinced the reader that scientists do typically aim at the most promising theory, a further question occurs. Should they do so? In other words, a defence from the normative aspect is necessary for a complete argument for promisingness as the criterion of theory choice.

McMullin (1976, 401) argues that the unit of appraisal in theory choice should not be a "timeless" entity. Rather it should be construed as something "developed and modified over a period of time." For McMullin, "a good theory is one which successfully guided research over an extended period of time, not only in the sense of providing correct predictions (an ad hoc theory could do this) but also in the sense of imaginatively suggesting modifications in the original theory/model, modifications that allowed predictions to be made which would not have been directly deducible from the original hypothesis." I agree with McMullin on the point that it is not wise to assess theories at a given time. For example, in theory appraisal, it is not sufficient to ask whether a theory T_1 makes more novel predictions than its rival T_2, or whether T_1 explain the phenomena better than T_2 at a given time *t*. T_1 and T_2 should be compared and judged over a period of time, say, *t* (in the past) to *t'* (now).

However, I wish to take a step forward. I argue that it is not sufficient to assess theories by examining its virtues in a period of time up to now (i.e. P-fertility). It seems to me that a diachronic criterion only concerning the past and the present is neither sufficient nor necessary to be a criterion of theory choice. First, if a theory T_1 is shown to be superior to another T_2 in the past and at present, it does not imply T_1 will be superior to T_2 after further development. Clearly, what scientists should aim at is not just a so-far best theory. Rather, in theory choice, scientist should choose the theory which will still be superior to others after further development in the future. Second, if a theory T_1 will be superior to its rival T_2 after further development, it does not imply that T_1 is shown to be superior to another T_2 in the past and at present. For example, the Copernican theory, when just introduced, was not as *P*-fertile as the Ptolemaic theory, which successfully made modifications in its original model in the past 1000 years. Nevertheless, Galileo and Kepler believed in the potential usefulness of the Copernican theory. Therefore, I argue that the promisingness criterion is better than the goodness criterion, from a normative point of view.

10.6 Conclusion

In a nutshell, I have argued that theory choice is a situation where scientists are reasoning whether a theory should be favoured over another as the more promising theory in the field rather than the one where scientists choose a theory among alternatives to be the best theory in the field. Accordingly, I have suggested that a philosophical analysis of theory choice should encompass an assessment of the promisingness of theories. I also have defined promisingness in terms of potential usefulness and have proposed that simplicity and external consistency typically project promisingness. Moreover, I have argued that my promisingness criterion is better than other diachronic criteria, including Popper's potential progressiveness, Lakatos' predictive novelty, McMullin's P-fertility, Laudan's fertility-promise, Ivani's fruitfulness, and Šešelja et al's pursuit worthiness. Finally, I have shown that the promisingness criterion is not only descriptively adequate but also normatively defensible.

References

Baker, Alan. 2003. Quantitative Parsimony and Explanatory Power. *The British Journal for the Philosophy of Science* 54 (2): 245–259.

Bateson, William. 1894. *Materials for the Study of Variation Treated with Especial Regard to Discontinuity in the Origin of Species*. London: Macmillan & Company.

———. 1902. *Mendel's Principles of Heredity: A Defence*. Cambridge: Cambridge University Press.

———. 1905. Compulsory Greek at Cambridge. *Nature* 71 (1843): 390.

———. 1909. *Mendel's Principles of Heredity*. Cambridge: Cambridge University Press.

Baumann, Peter. 2005. Theory Choice and the Intransitivity of 'Is a Better Theory Than'. *Philosophy of Science* 72 (1): 231–240.

Bradley, Seamus. 2017. Constraints on Rational Theory Choice. *The British Journal for the Philosophy of Science* 68 (3): 639–661.

Chang, Hasok. 2014. Epistemic Activities and Systems of Practice: Units of Analysis in Philosophy of Science After the Practical Turn. In *Science After the Practice Turn in the Philosophy, History and Social Studies of Science*, ed. Léna Soler, Sjoerd Zwart, Michael Lynch, and Vincent Israel-Jost, 67–79. New York/London: Routledge.

Correns, Carl. 1900. G. Mendels Regel über das Verhalten der Nachkommenschaft der Rassenbastarde. *Berichte der Deutschen Botanischen Gesellschaft* 18 (4): 158–168.

Darden, Lindley. 1977. William Bateson and the Promise of Mendelism. *Journal of the History of Biology* 10 (1): 87–106.

Forster, Malcolm, and Elliott Sober. 1994. How to Tell When Simpler, More Unified, or Less Ad Hoc Theories will Provide More Accurate Predictions. *The British Journal for the Philosophy of Science* 45 (1): 1–35.

Froggatt, Peter, and Norman C. Nevin. 1971. The 'Law of Ancestral Heredity' and the Mendelian-Ancestrian Controversy in England, 1889-1906. *Journal of Medical Genetics* 8 (1): 1–36.

Galton, Francis. 1889. *Natural Inheritance*. London/New York: Macmillan & Company.

Ivani, Silvia. 2019. What We (Should) Talk about When We Talk about Fruitfulness. *European Journal for Philosophy of Science* 9 (1): 4.

Ivanova, Milena, and Cedric Paternotte. 2013. Theory Choice, Good Sense and Social Consensus. *Erkenntnis* 78 (5): 1109–1132.

Kuhn, Thomas Samuel. 1957. *The Copernican Revolution*. Cambridge, MA: Harvard University Press.

———. 1970. *The Structure of Scientific Revolutions*. 2nd ed. Chicago, IL: The University of Chicago Press.

———. 1977. Objectivity, Value Judgment, and Theory Choice. In *The Essential Tension: Selected Studies in Scientific Tradition and Change*, 320–339. Chicago, IL: The University of Chicago Press.

———. 1983. Rationality and Theory Choice. *Journal of Philosophy* 80 (10): 563–570.
Lakatos, Imre. 1968. Criticism and the Methodology of Scientific Research Programmes. *Proceedings of the Aristotelian Society* 69: 149–186.
———. 1978. Falsification and the Methodology of Scientific Research Programmes. In *The Methodology of Scientific Research Programme*, ed. John Worrall and Gregory Currie, 8–101. Cambridge: Cambridge University Press.
Laudan, Larry. 1977. *Progress and Its Problems: Toward a Theory of Scientific Growth*. Berkeley/Los Angeles: The University of California Press.
———. 1981. A Problem-Solving Approach to Scientific Progress. In *Scientific Revolutions*, edited by Ian Hacking, 144–155. Oxford: Oxford University Press.
Mayr, Ernst. 1982. *The Growth of Biological Thought*. Cambridge, MA: The Belknap Press.
McMullin, Ernan. 1976. The Fertility of Theory and the Unit for Appraisal in Science. In *Essays in Memory of Imre Lakatos*, Boston Studies in the Philosophy of Science, ed. Robert S. Cohen, Paul Feyerabend, and M.W. Wartofsky, 395–432. Dordrecht/Boston: D. Reidel Publishing Company.
———. 1982. Values in Science Volume. *PSA: Proceedings of the Biennial Meeting of the Philosophy of Science Association* 2: 3–28.
———. 2014. The Virtues of a Good Theory. In *The Routledge Companion to Philosophy of Science*, ed. Martin Curd and Stathis Psillos, 2nd ed., 561–571. London/New York: Routledge.
Mendel, Gregor. 1866. Versuche über Pflanzenhybriden. *Verhandlungen des Naturforschenden Vereins Brünn* IV (Abhandlungen): 3–47.
Morreau, Michael. 2014. Mr. Fit, Mr. Simplicity and Mr. Scope: From Social Choice to Theory Choice. *Erkenntnis* 79 (6): 1253–1268.
———. 2015. Theory Choice and Social Choice: Kuhn Vindicated. *Mind* 124 (493): 239–262.
Norton, Bernard J. 1973. The Biometric Defense of Darwinism. *Journal of the History of Biology* 6 (2): 283–316.
Okasha, Samir. 2011. Theory Choice and Social Choice: Kuhn versus Arrow. *Mind* 120 (477): 83–115.
Olby, Robert Cecil. 1989. The Dimensions of Scientific Discovery: The Biometric-Mendelian Debate. *The British Journal for the History of Science* 22 (3): 299–320.
Pearson, Karl. 1898. Mathematical Contributions to the Theory of Evolution. On the Law of Ancestral Heredity. *Proceedings of the Royal Society of London* 62: 386–412.
———. 1908. On a Mathematical Theory of Determinantal Inheritance, from Suggestions and Notes of the Late W. F. R. Weldon. *Biometrika* 6 (1): 80–93.
Pearson, Karl, and Alice Lee. 1900. Mathematical Contributions to the Theory of Evolution. VIII. On the Inheritance of Characters Not Capable of Exact Quantitative Measurement. Part I. Introductory. Part II. On the Inheritance of Coat-Colour in Horses. Part III. On the Inheritance of Eye-Co. *Philosophical Transactions of the Royal Society of London. Series A, Containing Papers of a Mathematical or Physical Character* 195: 79–150.
Popper, Karl. 1963. Truth, Rationality, and the Growth of Scientific Knowledge. In *Conjectures and Refutations: The Growth of Scientific Knowledge*, 215–250. London: Routledge and Kegan Paul.
Provine, William B. 1971. *The Origin of Theoretical Population Genetics*. Chicago, IL: The University of Chicago Press.
Punnett, Reginald Crundall. 1911. *Mendelism*. 3rd ed. New York: The MacMillan Company.
Rizza, Davide. 2014. Arrow's Theorem and Theory Choice. *Synthese* 191 (8): 1–10.
Rueger, Alexander. 1996. Risk and Diversification in Theory Choice. *Synthese* 109 (2): 263–280.
Šešelja, Dunja, and Erik Weber. 2012. Rationality and Irrationality in the History of Continental Drift: Was the Hypothesis of Continental Drift Worthy of Pursuit? *Studies in History and Philosophy of Science* 43 (1): 147–159.
Šešelja, Dunja, Laszlo Kosolosky, and Christian Straßer. 2012. Rationality of Scientic Reasoning in the Context of Pursuit: Drawing Appropriate Distinctions. *Philosophica* 86: 51–82.

Waters, C.Kenneth. 2004. What was Classical Genetics? *Studies in History and Philosophy of Science* 35 (4): 783–809.

———. 2007. The Nature and Context of Exploratory Experimentation: An Introduction to Three Case Studies of Exploratory Research. *History and Philosophy of the Life Sciences* 29 (3): 275–284.

———. 2014. Shifting Attention from Theory to Practice in Philosophy of Biology. In *New Directions in the Philosophy of Science*, ed. Maria Carla Galavotti, Dennis Dieks, Wenceslao J. Gonzalez, Stephan Hartmann, Thomas Uebel, and Marcel Weber, 121–139. Dordrecht: Springer.

Weldon, Walter Frank Rapheal. 1894. The Study of Animal Variation. *Nature* 50: 25–26.

———. 1895. Remarks on Variation in Animals and Plants. To Accompany the First Report of the Committee for Conducting Statistical Inquiries into the Measurable Characteristics of Plants and Animals. *Proceedings of the Royal Society of London* 57: 379–382.

———. 1902a. Mendel's Laws of Alternative Inheritance in Peas. *Biometrika* 1 (2): 228–254.

———. 1902b. On the Ambiguity of Mendel's Categories. *Biometrika* 2 (1): 44–55.

Weldon, Walter Frank Rapheal, Karl Pearson, and Charles B. Davenport. 1901. Editorial: The Spirit of Biometrika. *Biometrika* 1 (1): 3–6.

Appendix

Evelyn Stern's Inconsistent Translation of *Eigenschaften* (de Vries 1966)

Original Text	Translation
. . ., so sind sie in diesen Eigenschaften antagonistisch, . . . (de Vries 1900, 84)	. . ., in these *characteristics* they are antagonistic, . . . (de Vries 1966, 109–10)
Der Kreuzungsversuch wird dadurch auf die antagonistisch Eigenschaften beschränkt. (de Vries 1900, 84)	The crossing experiment is thereby limited to the antagonistic *characteristics*. (de Vries 1966, 110)
Von den beiden antagonistischen Eigenschaften trägt der Bastard stets nur die eine, . . . (de Vries 1900, 84)	Of the two antagonistic *characteristics*, the hybrid carries only one, . . . (de Vries 1966, 110)
Bei der Bildung des Pollens und der Eizellen trennen sich die beiden antagonistischen Eigenschaften. (de Vries 1900, 84)	In the formation of pollen and ovules the two antagonistic *characteristics* separate, . . . (de Vries 1966, 110)
Das Fehlen von Mittelbildungen zwischen je zwei einfachen antagonistichen Eigenschaften im Bastard ist vielleicht der beste Beweis dafür, dass solche Eigenschaften wohl abgegrenzte Einheiten. (de Vries 1900, 85)	The lack of transitional forms between any two simple antagonistic *characters* in the hybrid is perhaps the best proof that such *characters* are well delimited units. (de Vries 1966, 110)
Von den beiden antagonistischen Eigenschaften (de Vries 1900, 85)	Of the two antagonistic *characters*, . . . (de Vries 1966, 111)
Gewöhnlich ist die systematisch höhere Eigenschaft die dominirende, oder bei bekannter Abstammung die ältere, . . . (de Vries 1900, 85)	Ordinarily the *character* higher in the systematic order is the dominating one, or, in cases of known ancestry, it is the older one. (de Vries 1966, 111)
IM Bastard liegen die beiden antagonistischen Eigenschaften als Anlagen neben einander. (de Vries 1900, 86)	In the hybrid the two antagonistic *characters* lie next to each other as anlagen. (de Vries 1966, 108–10)

(continued)

Y. Shan, *Doing Integrated History and Philosophy of Science: A Case Study of the Origin of Genetics*, Boston Studies in the Philosophy and History of Science 320,
https://doi.org/10.1007/978-3-030-50617-9

Original Text	Translation
Bei der Bildung der Pollenkörner und Eizellen trennen sie sich. Die einzelnen Paare antagonistischer Eigenschaften verhalten sich dabei unabhängig von einander. (de Vries 1900, 86)	In the formation of pollen grains and ovules these *characters* separate. The individual pairs of antagonistic *characters* behave independently during this process. (de Vries 1966, 112)
Die Individuen d und d^2 haben nur die dominirende, die Exemplare r und r^2 nur die recessive Eigenschaft, während die dr offenbar Bastarde sind. (de Vries 1900, 86)	The individuals d and d^2 have only the dominating *character*, those of r and r^2 constitution possess only the recessive *character*, while the dr plants are obviously hybrid. (de Vries 1966, 112)
Nennt man z. B. A. das eine, und B das andere Paar antagonistischer Eigenschaften, . . . (de Vries 1900, 89)	If, for instance, one pair of antagonistic *characters* is called A and the other pair B, . . . (de Vries 1966, 116)
Wendet man ferner den Satz an, dass die Bastarde das dominirende Merkmal zur Schau tragen, so findet man für die sichtbaren Eigenschaften der Nachkommenschaft. (de Vries 1900, 89)	If one applies the rule that hybrids exhibit the dominating traits, one finds for the visible *characteristics* of the progeny. (de Vries 1966, 116)
Es gelingt häufig, durch die Spaltungsveruche einfache Eigenschaften in mehrere Factoren zu zerlegen. (de Vries 1900, 89)	Success is frequently had in separating simple *characters* into a number of factors by means of segregation. (de Vries 1966, 117)

Index

A

Achinstein, P., 160–166, 172
Activeness, 43–46, 143
Analytic approach, 135–137, 141
Anlage, 89, 104, 140
Anti-anachronistic reading, 6–8

B

Bateson, W., 7, 27, 36, 43, 46, 47, 51, 53, 58, 69, 72, 75–77, 79, 90–95, 100–105, 107, 111–113, 117, 118, 121, 122, 124–129, 140, 141, 143, 145, 149–151, 160, 177–184
Biometric approach, 51, 61, 182
Biometricians, 11, 52–54, 58, 62–65, 101, 127, 177–180, 182–184
Biometric theory, 52, 54, 58, 61, 62, 69, 177–183
Biometry, 7, 51–54, 64–65, 177
Bird, 1, 99, 106–108, 113
Brannigan, A., 8, 15, 25, 78, 79
Brigandt, I., 137, 145

C

Callender, L.A., 8, 16, 118
Causation, 63, 64, 91
Cell-type, 46, 104, 167
Chang, H., 4, 5, 110, 137, 164
Cognitive approach, 135–137
Concepts, 3, 6, 8, 9, 26, 27, 36, 40, 42, 46, 47, 60, 72, 74–77, 79, 82, 84, 85, 90–92, 102–104, 109, 110, 112, 113, 124, 126, 127, 135–146, 148–153, 165, 183–185
Conceptual change, 2, 8, 133, 135–138, 144–146, 149, 151–153
Conceptual continuity, 9, 135–153
Conceptualisation, 14, 25–27, 31, 38, 40, 81, 83–85, 91–93, 101, 109, 111, 112, 126, 127, 133, 143, 145, 148, 151, 165–167, 169, 180
Conceptual practice, 133, 136, 137, 144–146, 148, 149, 152, 153
Confirmation, 8, 25, 82, 86, 88, 89, 160, 163, 167, 170, 171
Confrontation model, 3–5
Contextualism, 170–171
Corcos, A.F., 8, 36, 43, 118
Correns, C., 7, 8, 25, 27, 35, 36, 69, 71, 72, 75–77, 85, 89–93, 100–102, 104, 105, 111, 113, 117, 118, 122, 124, 126–129, 140, 141, 143, 145, 149–151, 160, 161, 163, 167, 177

D

Darbishire, A.D., 56, 58, 64
Darden, L., 8, 39, 72, 73, 76, 77, 141, 143, 144, 177
Darwin, C., 81, 99, 100, 102–105, 111, 112, 121, 122, 125
Development, 2, 4, 8, 11, 14–24, 28–30, 37, 52, 57–61, 64, 69, 71, 72, 74–77, 79, 83–86, 89–92, 94, 95, 101–104, 111–113, 118,

Y. Shan, *Doing Integrated History and Philosophy of Science: A Case Study of the Origin of Genetics*, Boston Studies in the Philosophy and History of Science 320,
https://doi.org/10.1007/978-3-030-50617-9

119, 121–126, 133, 135, 137–139, 141, 144, 145, 149–152, 179, 180, 186, 187
Developmental series, 8, 14, 17, 22–31, 92, 122
Development of hybrids, 14–16, 19–25, 28, 42, 75, 76, 84, 92, 123, 126, 151
De vries, H., 7, 8, 11, 26, 27, 35–47, 69, 71, 72, 75–77, 85, 88, 90–94, 100–104, 111–113, 117–119, 124, 126–129, 133, 139–141, 143, 145, 149–151, 159–161, 163, 167, 180, 191–192
Dominance, 9, 26, 27, 39–41, 43–47, 55, 56, 59–61, 64, 69, 73, 77, 85, 90–92, 103, 113, 124, 126, 127, 138–145, 149–153, 181
Duhem, P.M.M., 159, 168

E
Eigenschaften, 44, 140, 191–192
Entwicklungs, 8, 11, 14–22, 31, 118, 121
Entwicklungsreihe, 8, 14, 16, 17, 22–24
Evidence, 9, 22, 30, 38, 54, 62, 64, 120, 133, 157–172, 177, 181
Evidential pluralism, 63, 64
Evidential practices, 133, 160, 164–167, 170–172
Exemplar-based approach, 9, 69, 72, 79–84, 94, 100, 114, 133
Exemplars, 27, 79–84, 94, 108
Exemplary practices, 81, 83–86, 88, 89, 92–95, 100–102, 104, 117, 124, 126–129, 133, 149, 151, 165, 167, 176, 179, 180, 185
Experimentation, 27, 31, 36, 57, 74, 83–85, 93, 101, 109–112, 125, 127, 145, 148, 149, 151–153, 159, 164, 165, 180

F
Falk, R., 28, 137
Feest, U., 137, 145
Fertility, 86, 180, 182–186
Frame-like structure, 136, 141

G
Galton, F., 51, 53, 54, 56, 58–62, 65, 100, 125, 177, 179, 180
Gap problem, 9, 133, 157–164, 166–170, 172
Gärtner, C.F., 14–24, 30, 31, 69, 78, 121, 123
Giere, R.N., 1–3, 164
Gliboff, S., 14, 16–18, 21, 22, 31, 36

H
History and philosophy of science, 1–7
Holistic approach, 138, 144–145, 153
Hybridisation, 7, 18–20, 26, 39, 40, 42, 85, 103–105, 127, 138, 139, 143, 150
Hybrids, 13–15, 17–28, 30, 31, 36–47, 58, 61, 69, 71, 73, 75–77, 84–92, 94, 101, 103, 118, 122–127, 139–141, 143, 149–151, 157, 158, 191, 192
Hypothesisation, 31, 81, 83–86, 91–93, 101, 109, 112, 125–127, 143, 145, 148, 151–153
Hypothetico-deductivism, 9, 157–172

I
Incorporation, 36, 37, 46, 143
Integrated HPS, 4–9, 69, 71, 72, 133
Internalism, 107, 108
Irrelevant conjunctions, 159, 167–169

K
Kitcher, P., 72, 74, 76, 77
Know-how, 110, 111
Know-that, 11, 110, 168
Kuhn, T., 1, 78–80, 99, 101, 105–109, 145, 176, 178, 180

L
Lakatos, I, 1, 83, 105, 176, 183, 188
Laudan, L., 1, 2, 99, 101, 105–109, 176, 184, 188
Law of ancestral heredity, 51, 58, 61, 62, 65, 100, 177, 178, 180
Law of composition of hybrid fertilising cells, 9, 24, 46, 85, 91, 103, 111, 124, 125, 157
Law of segregation, 37–42, 45, 46, 60, 69, 73, 76, 77, 100, 104, 105, 110, 111, 118, 122, 127, 139, 141, 150, 151, 160, 161, 163, 167

M
McMullin, E., 176, 183, 184, 187, 188
Mendelian approach, 36, 37, 47, 51, 52, 61, 101, 104, 105, 110, 112, 125, 127, 143, 149
Mendelian ratios, 25, 27, 38–40, 104
Mendelian theory, 7, 36, 43, 52, 58–61, 64, 69, 103, 141, 176–181, 183, 184

Mendelism, 25, 35–48, 51–56, 60, 61, 64–65, 118, 125, 127, 129, 140, 177, 181
Mendels, G., 7–9, 11, 13–32, 35–47, 51, 53, 56, 58, 59, 61, 69, 71–73, 75–79, 84–86, 89–95, 101–104, 107, 108, 111–113, 117–129, 133, 138–141, 143–145, 149–151, 157–160, 163, 167, 172, 177, 180
Mendel's law, 28, 37, 40, 75, 77, 91, 111, 118, 151
Merkmal, 43, 44, 139, 140, 192
Monaghan, F.V., 8, 36, 43, 118
Morgan, T.H., 8, 61, 72, 74, 76, 77, 81, 92, 110, 141, 143–145, 149, 151, 160, 161
Müller-Wille, S., 8, 14–16, 22, 24, 77–79, 85, 89

N
Nägeli, C.W., 18, 25, 30, 31
Napp, F.C., 29, 30
Nickles, T., 5, 79, 84
Normativity, 170–171, 187

O
Olby, R.C., 8, 14–16, 28, 31, 36, 53, 54, 59, 64, 85, 91, 92, 117, 180, 181
Orel, V., 8, 14, 15, 22, 24, 28–31, 36, 77–79, 85, 89, 117, 118

P
Pangenesis, 36, 37, 40, 42–45, 76, 85, 91, 99, 100, 102–104, 112, 139, 140, 143
Pearson, K., 7, 52–54, 57, 58, 61–65, 127, 177–183
Pence, C.H., 8, 52, 63
P-fertility, 176, 183, 184, 187, 188
Pluralism, 63, 64
Popper, K., 176, 182, 183, 188
Practice-based approach, 135, 137, 145
Predictive novelty, 176, 183, 188
Problem-defining, 80, 81, 83–85, 93, 100–102, 105, 108–114, 117, 125, 126, 133, 145, 148, 149, 151–153, 179, 180, 184–186
Problem-solving, 81, 93, 99–102, 105–114, 117, 133, 145, 179, 180, 184, 185
Problem-specification, 81, 85, 89, 90, 92, 109, 124, 127, 143
Promisingness, 133, 175–188
Putnam, H., 136

R
Radick, G., 8, 52–54, 59
Reasoning, 74, 76, 82–86, 88, 89, 93, 109, 145, 148, 152, 153, 162, 165, 180, 187
Rediscovery, 7, 8, 11, 35–38, 71, 72, 117, 118, 125
Referential continuity, 136, 141

S
Sandlers, I., 14, 22, 28, 94, 121, 122
Sandlers, L., 94, 121, 122
Sarkar, S., 28
Schickore, J., 3, 4
Scientific concepts, 135–137, 144, 145, 149, 152, 153
Scientific progresses, 8, 99–101, 105–114, 184
Stamhuis, I.H., 8, 36, 38–40

T
Theorising, 42, 111
Theory choice, 9, 133, 165, 168, 170, 175–188

U
U-fertility, 184
Underdetermination, 159, 160, 169–170
Unger, F., 16, 17, 21, 22
Usefulness, 84, 85, 100–102, 104, 105, 108, 110–114, 117, 126–129, 133, 179–184, 186–188

V
von Tschermak, E., 7, 8, 27, 35, 36, 69, 71, 75–77, 90, 92, 93, 104, 117, 118, 124, 126–128, 141

W
Waters, C.K., 74, 77, 137, 161, 164, 186
Weldon, W.F.R., 7, 8, 11, 51–65, 69, 77, 101, 127, 177, 178, 181–183

Y
Yule, G.U., 64

The manufacturer's authorised representative in the EU is Springer Nature Customer Service Centre GmbH, Europaplatz 3, 69115 Heidelberg, Germany. If you have any concerns regarding our products, please contact ProductSafety@springernature.com

Printed and bound by CPI Group (UK) Ltd, Croydon, CR0 4YY
22/09/2026
02221058-0007